REFERENCE MATERIAL
FROM PLANE GEOMETRY

1. If a whole angle (a complete revolution) is divided into 360 equal parts, each of these is said to be an angle of *one degree*, denoted 1°.

2. Two angles are said to be *complementary* if their sum is 90°.

3. Two angles are said to be *supplementary* if their sum is 180°.

4. The sum of the three angles of a triangle is 180°.

5. In a right triangle, the sides that form the right angle are called the *legs;* the side opposite the right angle is called the *hypotenuse*.

6. **Pythagorean theorem:** *The square of the hypotenuse of a right triangle equals the sum of the squares of its legs.*

7. In a right triangle with angles 30°, 60°, 90°, the hypotenuse is twice the shorter leg.

8. An *isosceles* triangle has two equal sides, and hence two equal angles.

9. An *equilateral* triangle has three equal sides. Each angle is 60°.

10. A *tangent* line to a circle is perpendicular to the radius drawn to the point of tangency.

11. If two triangles have the three angles of one equal respectively to the three angles of the other, the triangles are *similar* and their corresponding sides are *proportional*.

PLANE
TRIGONOMETRY

FIFTH EDITION

PLANE TRIGONOMETRY

E. RICHARD HEINEMAN

TEXAS TECH UNIVERSITY

McGRAW-HILL BOOK COMPANY

New York St. Louis San Francisco Auckland Bogotá Hamburg
Johannesburg London Madrid Mexico Montreal New Delhi Panama
Paris São Paulo Singapore Sydney Tokyo Toronto

PLANE TRIGONOMETRY

567890   VHVH   898765432

Library of Congress Cataloging in Publication Data

Heineman, Ellis Richard.
 Plane trigonometry.

 Previous editions published under title: Plane trigonometry with tables.
 Includes index.
 1. Trigonometry, Plane. I. Title.
QA533.H47 1980 516'.24   79-19394
ISBN 0-07-027932-2

This book was set in Baskerville by York Graphic Services, Inc.
The editors were Carol Napier, Sibyl Golden, and James W. Bradley;
the designer was Jo Jones;
the production supervisor was Phil Galea.
New drawings were done by J & R Services, Inc.
Von Hoffmann Press, Inc., was printer and binder.

Cover photograph: © 1979 Barrie Rokeach.

CONTENTS

PREFACE

This fifth edition of *Plane Trigonometry* continues the emphasis of the third and fourth editions on the analytical aspects of the subject. In addition, the following changes have been made.

1. The text gives a detailed explanation of the use of a calculator in each type of problem involving computation. Logarithmic computation is mentioned in Chapter 11 and the Appendix.
2. Logarithmic and exponential functions, the properties of logarithms, and logarithmic equations are discussed in Chapter 10. The computational capabilities of logarithms are treated in the Appendix.
3. The wrapping function is introduced, and the treatment of the circular functions has been revised.
4. The Instructor's Manual contains (*a*) teaching aids and suggestions, (*b*) sample tests, (*c*) two comprehensive examinations, (*d*) answers to the tests and exams, (*e*) a discussion of the advantage of using a few unannounced quizzes (5 to 8 minutes in length), and (*f*) the answers to the text problems numbered 4, 8, 12, etc.

Instructors can save time in selecting problems by noticing the following features.

5. The problems in each exercise are so arranged that by assigning numbers 1, 5, 9, etc., or similar sets beginning with 2, 3, or 4, the instructor can obtain balanced coverage of all points involved without undue emphasis on some principles at the expense of others. For example, in the solution of right triangles in Exercise 11, each of the four sets of problems includes problems involving the use of the sine, the cosine, and the tangent; each set contains problems involving the angle of elevation, the angle of depression, and the concept of bearing; and each set contains approximately the same number of problems in which the unknown is an acute angle (or a leg, or the hypotenuse). This does not mean that the exercises consist of sets of four problems that are identical

except for numerical quantities. Wherever possible, the author has tried to make each problem different in some way, other than numerically, from all other problems in that exercise.

6. Answers to three-fourths of the problems are given at the back of the book. Answers to problems numbered 4, 8, 12, etc., appear in the Instructor's Manual.

The problem lists in this edition are completely new. Additional features retained from the fourth edition (1974) include the following:

7. Many of the exercises contain true-false questions to test the student's ability to avoid pitfalls and to detect camouflaged truths. The duty of the instructor is not only to teach correct methods but also to convince the student of the error in the false methods.

8. Definite instructions are given for proving identities and solving trigonometric equations. The subject of identities is approached gradually with practice in algebraic operations with the trigonometric functions.

9. A careful explanation of approximations and significant figures is given early in the text. The principle of accuracy in figures is adhered to throughout the book.

10. All problem sets are carefully graded and contain an abundance of simple problems that involve nothing more than the principles being discussed. There is also an ample supply of problems of medium difficulty and some "head-scratchers."

11. In addition to a discussion (Chapter 6) of the graphs of the trigonometric functions of θ, there is a body of material (Chapter 9) on graphical methods. Included are the graphs of $a \sin (bx + c)$, $a \sin x + b \cos x$, and $\sin^n x$. The two chapters on graphing are intentionally separated by Chapter 7 (Functions of Two Angles) and Chapter 8 (Trigonometric Equations), in the belief that this arrangement will result in better comprehension by the student.

12. Miscellaneous points include (*a*) a note to the student, (*b*) problems that are encountered in calculus, (*c*) a careful explanation of the concept of infinity, (*d*) memory schemes, (*e*) the uses of the sine and cosine curves, and (*f*) interesting applied problems.

The author gratefully acknowledges Professor Robert C. Gebhardt, County College of Morris; Professor Frank Kocher, The Pennsylvania State University; Kathryn McClellan, Tulsa Junior College; Ara B. Sullenberger, Tarrant County Junior College; and Professor Wesley W. Tom, Chaffey College, for reading the manuscript and making valuable suggestions.

E. Richard Heineman

NOTE TO THE STUDENT

A mastery of the subject of trigonometry requires (1) a certain amount of memory work, (2) a great deal of practice and drill in order to acquire experience and skill in the application of the memory work, and (3) an insight and understanding of "what it is all about." Your instructor is a "troubleshooter" who attempts to prevent you from going astray, supplies missing links in your mathematical background, and tries to indicate the "common sense" approach to the problem.

The memory work in any course is one thing that students can and should perform by themselves. The least you can do for your instructor and yourself is to *commit to memory each definition and theorem as soon as you contact it.* This can be accomplished most rapidly not by reading, but by writing the definition or theorem until you can reproduce it without the aid of the text.

In working the problems, do not continually refer back to the illustrative examples. Study the examples so thoroughly (by writing them) that you can reproduce them with your text closed. Only after the examples are entirely clear and have been completely mastered should you attempt the unsolved problems. These problems should be worked *without referring to the text.*

Bear in mind, too, that *memory* and technical *skill* are aided by *understanding;* therefore, as the course develops you should review the definitions and theorems from time to time, always seeking a deeper insight into them.

E. Richard Heineman

PLANE TRIGONOMETRY

1
THE TRIGONOMETRIC FUNCTIONS

1 TRIGONOMETRY

Trigonometry is that branch of mathematics which deals primarily with six ratios called the trigonometric functions. These ratios are important for two reasons. First, they are the basis of a theory which is used in other branches of mathematics as well as in physics and engineering. Second, they are used in solving triangles. From geometry we recall that two sides and the included angle of a triangle suffice to fix its size and shape. It will be shown later that the length of the third side and the size of the remaining angles can be computed by means of trigonometry.

2 DIRECTED SEGMENTS

A directed line is a line upon which one direction is considered positive; the other, negative. Thus in Figure 1 the arrowhead indicates that all segments measured from left to right are positive. Hence if $OA = 1$ unit of length, then $OB = 3$, and $BC = -5$. Observe that since the line is directed, CB is not equal to BC. However, $BC = -CB$; or $CB = -BC$. Also note that $OB + BC + CO = 0$.

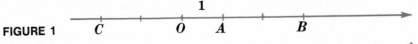

FIGURE 1

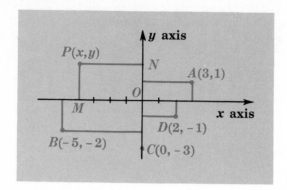

FIGURE 2

3 THE RECTANGULAR COORDINATE SYSTEM

A rectangular (or cartesian) coordinate system consists of two perpendicular *directed* lines. It is conventional to draw and direct these lines as in Figure 2. The *x* axis and the *y* axis are called the **coordinate axes;** their intersection O is called the **origin.** The position of any point in the plane is fixed by its distances from the axes.

> The *x* coordinate* (or *x*) of point *P* is the directed segment *NP* (or *OM*) measured from the *y* axis to point *P*. The *y* coordinate* (or *y*) of point *P* is the directed segment *MP*, measured from the *x* axis to point *P*.

It is necessary to remember that each coordinate is measured *from axis to point.* Thus the *x* of *P* is *NP* (not *PN*); the *y* of *P* is *MP* (not *PM*). The point *P*, with *x* coordinate *x* and *y* coordinate *y*, is denoted by $P(x, y)$. It follows that the *x* of any point to the right of the *y* axis is positive; to the left, negative. Also the *y* of any point above the *x* axis is positive; below, negative.

To **plot** a point means to locate and indicate its position on a coordinate system. Several points are plotted in Figure 2.

The distance from the origin O to point P is called the **radius vector** (or *r*) of *P*. This distance *r* is not directed and *is always positive*† by agreement. Hence with each point of the plane we can associate three coordinates: *x*, *y*, and *r*. The radius vector *r* can be found by using the pythagorean‡ relation $x^2 + y^2 = r^2$ (see Figure 3).

---

*The *x* coordinate and *y* coordinate are also called the *abscissa* and *ordinate,* respectively.

† Or zero (for the origin *O*).

‡ *Pythagorean theorem: The square of the hypotenuse of a right triangle equals the sum of the squares of its legs.*

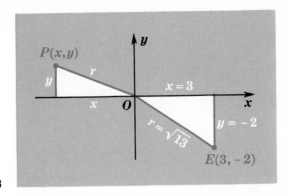

FIGURE 3

The coordinate axes divide the plane into four parts called **quadrants** as indicated in Figure 4. We shall sometimes denote these as Q I, Q II, Q III, and Q IV, respectively.

ILLUSTRATION 1   To find r for the point $(5, -12)$, use

$$r^2 = 5^2 + (-12)^2 = 169 \qquad r = 13$$

ILLUSTRATION 2   If $x = 15$ and $r = 17$, we obtain y by using

$$x^2 + y^2 = r^2$$

Hence $(15)^2 + y^2 = (17)^2$; $225 + y^2 = 289$; $y^2 = 64$; $y = \pm 8$. If the point is in quadrant I, $y = 8$; if the point is in quadrant IV, $y = -8$. (Since x is positive, the point cannot lie in either quadrant II or quadrant III.)

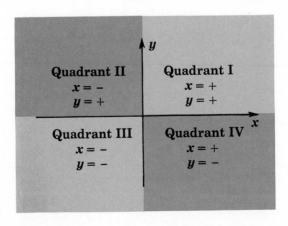

FIGURE 4

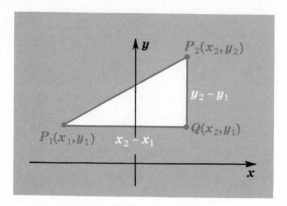

FIGURE 5

4 THE DISTANCE FORMULA

Let $P_1(x_1, y_1)$ and $P_2(x_2, y_2)$ be *any* two points in the xy plane (Figure 5). We shall use the pythagorean theorem to express the distance P_1P_2 in terms of the coordinates of the points. Through P_1, draw a line parallel to the x axis. Through P_2, draw a line parallel to the y axis. These lines meet at $Q(x_2, y_1)$. The length of the positive segment P_1Q is $x_2 - x_1$ (that is, the *right x* minus the *left x**). And the length of the positive segment QP_2 is $y_2 - y_1$ (that is, the *upper y* minus the *lower y*). Since $(P_1P_2)^2 = (P_1Q)^2 + (QP_2)^2$,

$$P_1P_2 = \sqrt{(x_2 - x_1)^2 + (y_2 - y_1)^2}$$

This formula holds for all positions of P_1 and P_2. If, for example, P_2 lies below P_1, then $QP_2 = y_2 - y_1$, which is a negative quantity. But the square of $(y_2 - y_1)$ is equal to the square of the positive quantity $(y_1 - y_2)$; that is, $(y_2 - y_1)^2 = (y_1 - y_2)^2$. Hence

the distance d between $P_1(\mathbf{x}_1, y_1)$ and $P_2(x_2, y_2)$ is

$$d = \sqrt{(x_2 - x_1)^2 + (y_2 - y_1)^2}$$

ILLUSTRATION   The distance between $A(-1, 5)$ and $B(3, -2)$ is $d = \sqrt{(-1 - 3)^2 + (5 + 2)^2} = \sqrt{16 + 49} = \sqrt{65}$. Either A or B can be designated as P_1.

---

*More precisely, the x of the point on the right (the large x) minus the x of the point on the left (the small x). For $P_1(-2, 1)$ and $Q(5, 1)$, $P_1Q = 5 - (-2) = 7$.

EXERCISE 1

1. Plot the following points on coordinate paper and then find the value of r for each: $(4, 3)$, $(0, -9)$, $(-3, 1)$, $(-2, -\sqrt{5})$.
2. Plot the following points on coordinate paper and then find the value of r for each: $(-7, -24)$, $(-6, 0)$, $(5, -2)$, $(\sqrt{7}, 3)$.
3. Plot the following points on coordinate paper and then find the value of r for each: $(15, -8)$, $(1, 0)$, $(6, 7)$, $(-2, \sqrt{21})$.
4. Plot the following points on coordinate paper and then find the value of r for each: $(-6, 8)$, $(0, 3)$, $(-4, -5)$, $(\sqrt{15}, -7)$.
5. Use the pythagorean theorem to find the missing coordinate and then plot the point:
   (a) $x = 24$, $r = 25$, point is in Q IV.
   (b) $y = 8$, $r = 9$, point is in Q I.
   (c) $x = -7$, $r = 7$.
6. Find the missing coordinate and then plot the point:
   (a) $y = 8$, $r = 17$, point is in Q II.
   (b) $x = -5$, $r = 2\sqrt{7}$, point is in Q III.
   (c) $x = 0$, $r = 3$, y is negative.
7. Find the missing coordinate and then plot the point:
   (a) $y = -4$, $r = 5$, point is in Q III.
   (b) $x = 2\sqrt{5}$, $r = 6$, point is in Q IV.
   (c) $y = 0$, $r = 2$, x is negative.
8. Find the missing coordinate and then plot the point:
   (a) $x = 12$, $r = 13$, point is in Q I.
   (b) $y = 7$, $r = 5\sqrt{2}$, point is in Q II.
   (c) $y = -1$, $r = 1$.
9. In which quadrants is the following ratio positive?

   (a) $\dfrac{y}{r}$ (b) $\dfrac{x}{r}$ (c) $\dfrac{y}{x}$

10. In which quadrants is the following ratio negative?

    (a) $\dfrac{y}{r}$ (b) $\dfrac{x}{r}$ (c) $\dfrac{y}{x}$

11. What is the y of all points on the x axis? What is the x of all points on the y axis?
12. Without plotting, identify the quadrant in which each of the following points lies if s is a negative number: $T(s, 4)$, $U(5, -s)$, $V(-s^2, s)$, $W(s^2, s^3)$.

13. Use the distance formula to find the exact value of the distance between the points $(-7, 1)$ and $(2, -4)$.

14. Use the distance formula to find the distance between the points $(3, -5)$ and $(-1, a)$.

5 TRIGONOMETRIC ANGLES

There are several ways of measuring angles. One of the oldest and most used methods is to divide one revolution into 360 equal parts; each part is then called one **degree** ($1°$).

In geometry, you may have thought of an angle as the "opening" between two rays* which form the sides of the angle and which emerge from a point called the vertex of the angle.

> A trigonometric angle is an amount of rotation used in moving a ray from one position to another.

A *positive* angle is generated by *counterclockwise* rotation; a *negative* angle, by *clockwise* rotation. Figure 6 illustrates the terms used and shows an angle of 200°. Figure 7 shows angles of 500° and $-420°$. The $-420°$ angle may be thought of as the amount of rotation effected by the minute hand of a clock between 12:15 and 1:25. *To specify a trigonometric angle, we need,* in addition to its sides, *a curved arrow extending from its initial side to its terminal side.*

---

*A *ray*, or half-line, is the part of a line extending in one direction from a point on the line.

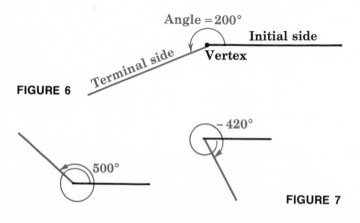

Angle = 200°

Initial side

Terminal side

Vertex

FIGURE 6

500°

$-420°$

FIGURE 7

6 STANDARD POSITION OF AN ANGLE

An angle is said to be in standard position if its vertex is at the origin and its initial side coincides with the positive x axis.

An angle is said to be in a certain quadrant if its terminal side lies in that quadrant *when the angle is in standard position*. For example, 600° is in the third quadrant; or, −70° is a fourth-quadrant angle.

Angles are said to be **coterminal** if their terminal sides coincide when the angles are in standard position. For example, 200°, 560°, and −160° are coterminal angles. From a trigonometric viewpoint these angles are not equal; they are merely coterminal.

EXERCISE 2

Place each of the following angles in standard position; draw a curved arrow to indicate the rotation. Draw and find the size of two other angles, one positive and one negative, that are coterminal with the given angle.

1. 130°	**2.** 320°	**3.** 40°	**4.** 230°
5. 450°	**6.** 380°	**7.** 460°	**8.** 710°

Each of the following points is on the terminal side of a positive angle in standard position. Plot the point; draw the terminal side of the angle; indicate the angle by a curved arrow; use a protractor to find, to the nearest degree, the size of the angle.

9. $(2, -10)$	**10.** $(-3, 8)$	**11.** $(-7, -2)$	**12.** $(8, 5)$
13. $(3, \sqrt{3})$	**14.** $(-4, -7)$	**15.** $(5, -9)$	**16.** $(-7, 6)$

17. A wheel makes 1300 revolutions per minute (rpm). Through how many degrees does it move in 1 second (sec)?

7 DEFINITIONS OF THE TRIGONOMETRIC FUNCTIONS OF A GENERAL ANGLE

The whole subject of trigonometry is based upon the six **trigonometric functions**. The names of these functions, with their abbreviations in parentheses, are: sine **(sin)**, cosine **(cos)**, tangent **(tan)**, cotangent **(cot)**, secant **(sec)**, cosecant **(csc)**. In a certain sense, the following definitions are the most important in this book.

A COMPLETE DEFINITION OF THE TRIGONOMETRIC FUNCTIONS OF ANY ANGLE θ

1. *Place the angle θ* in standard position. (See Figure 8.)*
2. *Choose any point P† on the terminal side of θ.*
3. *Drop a perpendicular from P to the x axis, thus forming a triangle of reference for θ.*
4. *The point P has three coordinates x, y, r, in terms of which we define the following trigonometric functions:*

$$a \quad \sin \theta = \frac{y}{r}$$

$$b \quad \cos \theta = \frac{x}{r}$$

$$c \quad \tan \theta = \frac{y}{x}$$

$$c' \quad \cot \theta = \frac{x}{y}$$

$$b \quad \sec \theta = \frac{r}{x}$$

$$a' \quad \csc \theta = \frac{r}{y}$$

---

*See Greek alphabet on front endpaper.

† Other than the origin O.

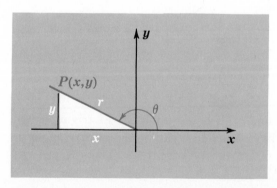

FIGURE 8

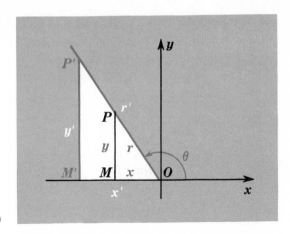

FIGURE 9

These six ratios are called *functions,* in accordance with the modern definition of the word, because they give rise to *ordered pairs*—for example, $(\theta, \sin\theta)$—with only one value ($\sin\theta$ in our example) corresponding to a given θ.

The *domain* of each function consists of all θ for which the corresponding denominator is not 0; thus $\tan\theta$ and $\sec\theta$ are not defined when $x = 0$, and $\cot\theta$ and $\csc\theta$ are not defined when $y = 0$. You will recall that *division by zero is impossible.* The definition of division states that $a/b = c$ if and only if $bc = a$, provided c is a unique number. If $\frac{1}{0} = a$, then $(0)(a)$ must equal 1. No such number a exists. Another explanation: When we write $\frac{12}{3}$, we are asking, "How many 3's add up to 12?" Consequently $\frac{1}{0}$ means "How many zeros will add up to 1?" Such a question is obviously absurd.

The *range* of each trigonometric function will be discussed later. The range of $\sin\theta$, for example, is by definition the set of all values taken on by $\sin\theta$ as θ varies over its domain.

While a function is defined to be a set of ordered pairs, such as $(\theta, \sin\theta)$, it is customary to speak somewhat loosely of the second member as a function of the first. Thus we say that $\sin\theta$ is a function of θ. In general, we say that *a function of θ is a quantity whose value is determined whenever a value in the domain of θ is assigned to θ.* For example, $3\theta^2 + 1$ is a function of θ for all numbers θ. If θ has the value 5, then $3\theta^2 + 1$ has the value 76. If $\theta = -4$, then $3\theta^2 + 1 = 49$. Likewise, $\theta^3 + 7$ and 8θ are functions of θ for all θ. Also, $(\theta - 2)/(\theta - 3)$ is a function of θ for all θ except $\theta = 3$.

In order to prove that $\sin\theta$ is a function of θ, we must show that the value of $\sin\theta$ is independent of the choice of point P on the terminal side of θ. Let P' (x', y') be any other point on OP (see Figure 9). Then, using the coordinates of P', we have $\sin \theta = y'/r'$. Since triangles $OP'M'$ and OPM are similar, it follows that

$$\frac{y'}{r'} = \frac{y}{r}$$

and the value of $\sin \theta$ is the same whether it is obtained by using P or by using P'. Since the value of $\sin \theta$ is determined by the value of θ and is independent of the choice of P, we can say that $\sin \theta$ is a function of θ. *The values of the trigonometric functions of θ depend solely upon the value of θ.*

8 CONSEQUENCES OF THE DEFINITIONS

(a) The following pairs of trigonometric functions are reciprocals of each other:

$$\sin \theta, \csc \theta \qquad \cos \theta, \sec \theta \qquad \tan \theta, \cot \theta$$

The reciprocal of the number a is $\dfrac{1}{a}$. Hence the reciprocal of 3 is $\frac{1}{3}$; the reciprocal of $-\frac{1}{4}$ is -4; $\frac{2}{5}$ and $\frac{5}{2}$ are reciprocals. Since $\dfrac{x}{y} = \dfrac{1}{y/x}$, we can say, for values of θ for which these functions are defined, that

$$\cot \theta = \frac{1}{\tan \theta}$$

Similarly,
$$\sec \theta = \frac{1}{\cos \theta}$$

and
$$\csc \theta = \frac{1}{\sin \theta}$$

Multiplying both sides of the last equation by $\sin \theta$, we get

$$\sin \theta \csc \theta = 1$$

Dividing both sides of this equation by $\csc \theta$, we obtain

$$\sin \theta = \frac{1}{\csc \theta}$$

Hence *sin θ* and *csc θ* are *reciprocals*; also *cos θ* and *sec θ* are *reciprocals*; and *tan θ* and *cot θ* are *reciprocals*. The following list indicates the reciprocal functions:

$$
\left.\begin{matrix}
\sin \theta \\
\cos \theta \\
\tan \theta \\
\cot \theta \\
\sec \theta \\
\csc \theta
\end{matrix}\right\} \quad \text{Reciprocals}
$$

Caution The symbol *cos* in itself has no meaning.* To have interpretation, it must be followed by some angle. Write *cos θ*, not *cos*. Notice that $\sin \theta \csc \theta = 1$ means that the sine of any angle times the cosecant of the *same* angle equals unity.

(b) Any trigonometric function of an angle is equal to the same function of all angles coterminal with it.

This follows directly from Section 7. Thus

$$\sin 370° = \sin (370° - 360°) = \sin 10°$$
$$\cos (-100°) = \cos (-100° + 360°) = \cos 260°$$
$$\tan 900° = \tan (900° - 720°) = \tan 180°$$

(c) The sine is positive for angles in the top quadrants; the cosine is positive for angles in the right-hand quadrants.

Since *r* is always positive, $\sin \theta$ is positive whenever *y* is positive, i.e., in the upper quadrants, I and II. Similarly $\sin \theta$ is negative in the lower quadrants, III and IV. Also $\sin \theta$ is 0 when $y = 0$, which occurs when θ is coterminal with 0° or 180° (see Figure 10).

Likewise, $\cos \theta$ has the same sign as *x*. Hence $\cos \theta$ is positive in the right-hand quadrants, I and IV. Also $\cos \theta$ is negative in the left-hand quadrants, II and III. And $\cos \theta$ is 0 whenever $x = 0$, which occurs when θ is coterminal with 90° or 270°.

---

*We can, however, speak of *the cosine function,* referring to the entire set of ordered pairs (θ, $\cos \theta$).

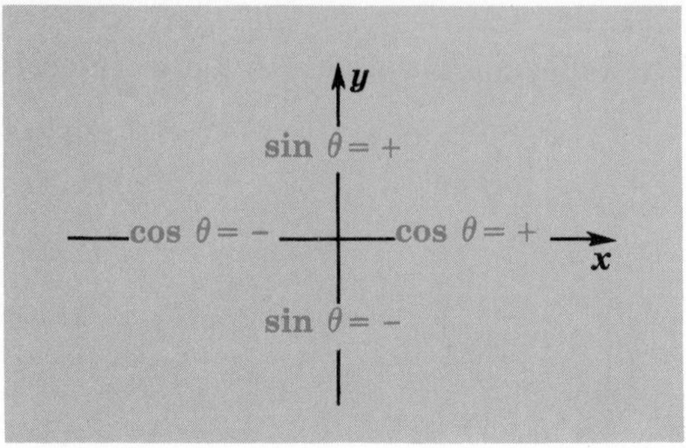

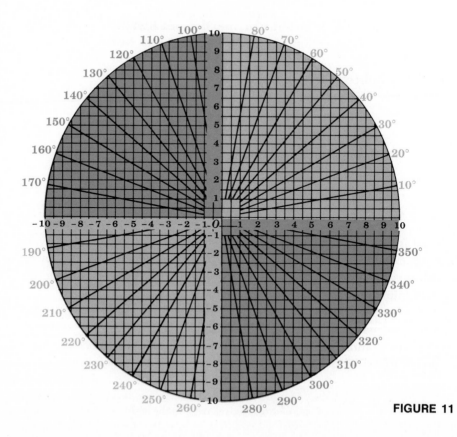

FIGURE 10

Moreover, $\tan \theta = y/x$ is positive when y and x have the same sign, namely in quadrants I and III. Also, $\tan \theta$ is negative when y and x have opposite signs, namely in quadrants II and IV.

The sign of each of the three remaining functions is the same as the sign of its reciprocal. Thus $\csc \theta$ is positive in the upper quadrants.

FIGURE 11

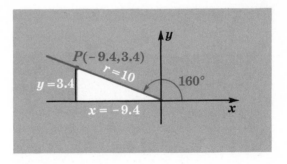

FIGURE 12

EXAMPLE 1 Find the values of sin 160°, cos 160°, tan 160°.

SOLUTION In Figure 11 the angle 160° is in standard position. For convenience, on the terminal side of 160° choose point P so that $r = 10$. In forming a triangle of reference we find that $x = -9.4$, $y = 3.4$* (see Figure 12). Hence

$$\sin 160° = \frac{y}{r} = \frac{3.4}{10} = 0.34$$

$$\cos 160° = \frac{x}{r} = \frac{-9.4}{10} = -0.94$$

$$\tan 160° = \frac{y}{x} = \frac{3.4}{-9.4} = -0.36$$

EXAMPLE 2 Find the trigonometric functions of 180°.

SOLUTION Place the angle in standard position. For point P let us choose $(-1, 0)$. Since r is always positive, $r = 1$. The triangle of reference has "collapsed," but P does have the coordinates x, y, r (see Figure 13). Then

$$\sin 180° = \frac{y}{r} = \frac{0}{1} = 0$$

$$\cos 180° = \frac{x}{r} = \frac{-1}{1} = -1$$

---

*Since this number was obtained from a drawing, it is merely an approximation. Equally acceptable would be 3.3 or 3.5.

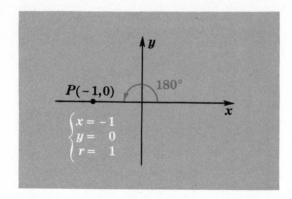

$$\begin{cases} x = -1 \\ y = \ 0 \\ r = \ 1 \end{cases}$$

FIGURE 13

$$\tan 180° = \frac{y}{x} = \frac{0}{-1} = 0$$

$$\cot 180° = \frac{x}{y} = \frac{-1}{0} \qquad \textit{which does not exist}$$

$$\sec 180° = \frac{r}{x} = \frac{1}{-1} = -1$$

$$\csc 180° = \frac{r}{y} = \frac{1}{0} \qquad \textit{which does not exist}$$

EXAMPLE 3 Assuming angle θ is in standard position, compute the trigonometric functions of θ if point $P(-2, -3)$ is on its terminal side.

SOLUTION The pythagorean theorem gives us $r = \sqrt{13}$. Angle θ and its triangle of reference are shown in Figure 14. Then

$$\sin \theta = \frac{y}{r} = \frac{-3}{\sqrt{13}} = -\frac{3\sqrt{13}*}{13}$$

$$\cos \theta = \frac{x}{r} = \frac{-2}{\sqrt{13}} = -\frac{2\sqrt{13}*}{13}$$

$$\tan \theta = \frac{y}{x} = \frac{-3}{-2} = \frac{3}{2}$$

---

*Rationalize the denominator by multiplying top and bottom by $\sqrt{13}$.

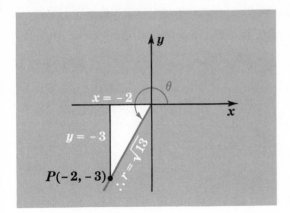

FIGURE 14

$$\cot \theta = \frac{x}{y} = \frac{-2}{-3} = \frac{2}{3}$$

$$\sec \theta = \frac{r}{x} = \frac{\sqrt{13}}{-2} = -\frac{\sqrt{13}}{2}$$

$$\csc \theta = \frac{r}{y} = \frac{\sqrt{13}}{-3} = -\frac{\sqrt{13}}{3}$$

EXERCISE 3

Place each of the following angles in standard position, using a curved arrow to indicate the rotation. Use Figure 11 to label the sides of the triangle of reference. Choose P so that r = 10. Then approximate the values of the sine, cosine, and tangent of each angle.

1. 230°
2. 20°
3. 170°
4. 320°
5. 100°
6. 280°
7. 250°
8. 70°
9. 350°
10. 130°
11. 40°
12. 200°

13. Using $r = 10$, read from Figure 11 the sine of each of the following angles: 0°, 10°, 20°, 30°, 40°, 50°, 60°, 70°, 80°, 90°.
14. Using $r = 10$, read from Figure 11 the cosine of each of the following angles: 0°, 10°, 20°, 30°, 40°, 50°, 60°, 70°, 80°, 90°.
15. Draw a figure and compute the trigonometric functions of 0°.
16. Draw a figure and compute the trigonometric functions of 90°.

17. Draw a figure and compute the trigonometric functions of 270°.

18. Draw a figure and compute the trigonometric functions of −180°.

Each of the following points is on the terminal side of an angle θ, in standard position. Use a curved arrow to specify θ. Construct and label the sides of the triangle of reference as in Figure 14. Find the exact values of the six trigonometric functions of θ. Leave the results in fractional form. Do not use a calculator.

19. $(24, -7)$ **20.** $(-8, 15)$ **21.** $(-12, -5)$ **22.** $(3, 4)$

23. $(-2, 9)$ **24.** $(7, -5)$ **25.** $(1, 4)$ **26.** $(-8, -1)$

27. $(-\sqrt{55}, -3)$ **28.** $(5, \sqrt{39})$ **29.** $(-\sqrt{3}, \sqrt{6})$ **30.** $(7, -4\sqrt{2})$

Copy the following statements and identify the quadrant in which θ must be in order to satisfy each set of conditions.

31. $\sin \theta = +$ and $\tan \theta = -$ **32.** $\cos \theta = -$ and $\csc \theta = -$

33. $\tan \theta = -$ and $\sec \theta = +$ **34.** $\tan \theta = -$ and $\csc \theta = +$

35. $\cot \theta = +$ and $\sec \theta = -$ **36.** $\sin \theta = -$ and $\cos \theta = +$

37. $\sec \theta = -$ and $\csc \theta = -$ **38.** $\cos \theta = +$ and $\tan \theta = -$

Copy the following statements, identify each as possible or impossible, and explain why. Do not use a calculator.

39. $\sin \theta = 0$ **40.** $\sin \theta = 1.5$

41. $\tan \theta = 0.01$ **42.** $\tan \theta = 3000$

43. $\cos \theta = 6$ **44.** $\cos \theta = 0$

45. $\sin \theta = -\frac{1}{5}$ and $\csc \theta = 5$ **46.** $\cot \theta = 7$ and $\csc \theta = 5$

Copy the following statements and in each case state whether θ is close to 0° or close to 90°. Do not use a calculator.

47. $\cos \theta = 0.001$ **48.** $\cos \theta = 0.999$

49. $\tan \theta = 0.01$ **50.** $\tan \theta = 100$

51. $\sin \theta = 0.01$ **52.** $\sin \theta = 0.99$

In Problems 53 to 56, refer to Figure 15; then fill in each blank with first, second, third, or fourth. (Review the explanation in Section 6 of how to determine the quadrant in which an angle lies.)

53. α is a _____-quadrant angle.   **54.** β is a _____-quadrant angle.

55. γ is a _____-quadrant angle.   **56.** δ is a _____-quadrant angle.

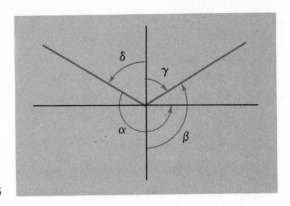

FIGURE 15

9 GIVEN ONE TRIGONOMETRIC FUNCTION OF AN ANGLE, TO DRAW THE ANGLE AND FIND THE OTHER FUNCTIONS

When we know (1) the quadrant in which an angle lies and (2) the value of one trigonometric function of this angle, it is possible, by using the pythagorean theorem and the general definition, to draw the angle and find its other five trigonometric functions.

EXAMPLE   Given $\cos \theta = \frac{2}{5}$ and θ is not in Q I, draw θ and find its other functions.

SOLUTION   Since $\cos \theta$ is positive in the two right-hand quadrants and since Q I is ruled out, θ must lie in Q IV. Remembering that for all angles $\cos \theta = x/r$ and that in this case $\cos \theta = \frac{2}{5}$, we can use $x = 2$ and $r = 5$.* By means of $x^2 + y^2 = r^2$, we find $y = -\sqrt{21}$, the negative sign being chosen because $P(x, y)$ is in Q IV (see Figure 16). Then

$$\sin \theta = \frac{y}{r} = \frac{-\sqrt{21}}{5} = -\frac{\sqrt{21}}{5}$$

$$\tan \theta = \frac{y}{x} = \frac{-\sqrt{21}}{2} = -\frac{\sqrt{21}}{2}$$

---

*Equally correct but not so convenient would be $x = 6$, $r = 15$ or $x = 1$, $r = \frac{5}{2}$.

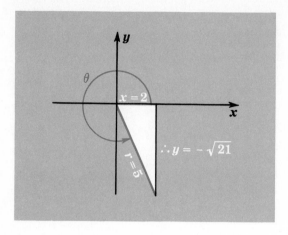

FIGURE 16

$$\cot \theta = \frac{x}{y} = \frac{2}{-\sqrt{21}} = -\frac{2\sqrt{21}}{21}$$

$$\sec \theta = \frac{r}{x} = \frac{5}{2}$$

$$\csc \theta = \frac{r}{y} = \frac{5}{-\sqrt{21}} = -\frac{5\sqrt{21}}{21}$$

EXERCISE 4

Construct and label the sides of the triangle of reference. Use a curved arrow to indicate θ. Find the exact values of the remaining trigonometric functions.

1. $\cos \theta = \frac{1}{7}$, θ not in Q I

2. $\tan \theta = -5$, θ not in Q IV

3. $\sin \theta = \frac{7}{8}$, θ not in Q II

4. $\cos \theta = -\frac{3}{4}$, θ not in Q II

5. $\tan \theta = \frac{15}{8}$, $\cos \theta > 0*$

6. $\sin \theta = -\frac{5}{13}$, $\tan \theta < 0$†

7. $\cos \theta = -\frac{40}{41}$, $\sin \theta > 0$

8. $\tan \theta = -\frac{24}{7}$, $\sin \theta < 0$

9. $\sin \theta = -\frac{2\sqrt{13}}{13}$, θ not in Q IV. *Hint:* $-\frac{2\sqrt{13}}{13} = \frac{-2}{\sqrt{13}}$

---

*The symbol $>$ is read "is greater than."

†The symbol $<$ is read "is less than."

10. $\cos \theta = \dfrac{\sqrt{77}}{9}$, θ not in Q IV

11. $\tan \theta = \dfrac{2\sqrt{5}}{15}$, θ not in Q I. *Hint:* $\dfrac{2\sqrt{5}}{15} = \dfrac{2\sqrt{5}}{3(5)} = \dfrac{-2}{-3\sqrt{5}}$

12. $\sin \theta = \dfrac{2\sqrt{10}}{7}$, θ not in Q I

13. $\csc \theta = 9$, θ not in Q I

14. $\cot \theta = 2$, θ not in Q I

15. $\sec \theta = b$, θ *is* in Q IV

16. $\cot \theta = a$, θ *is* in Q I

2 TRIGONOMETRIC FUNCTIONS OF AN ACUTE ANGLE

10 TRIGONOMETRIC FUNCTIONS OF AN ACUTE ANGLE

Let θ be an acute angle of a right triangle (Figure 17). If θ were in standard position, this right triangle would be its triangle of reference.* The hypotenuse (*hyp*) would be the *r* of some point *P* on the terminal side of θ; the side opposite θ (*opp*) would be the *y* of *P*; and the side adjacent θ (*adj*) would be the *x* of *P*. Then the general definitions (Section 7)

---

*In some cases it may be necessary to turn the triangle over.

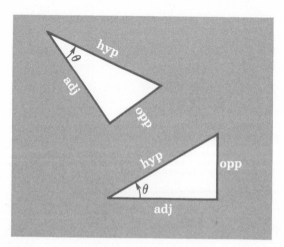

FIGURE 17

involving *x*, *y*, and *r* would become special definitions involving *adj*, *opp*, and *hyp*.

We conclude that *for any* **acute** *angle θ lying in a right triangle,*

$$\sin \theta = \frac{\text{opp}}{\text{hyp}}$$

$$\cos \theta = \frac{\text{adj}}{\text{hyp}}$$

$$\tan \theta = \frac{\text{opp}}{\text{adj}}$$

The other three functions can be obtained through their reciprocals.

EXERCISE 5

For each of the following right triangles, write the sine, cosine, and tangent of each acute angle. Leave results in fractional form.

1.   **2.**   **3.**   **4.**

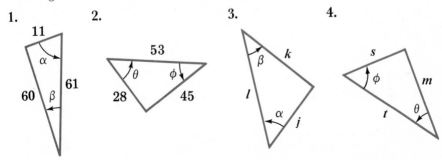

11 COFUNCTIONS

The sine and cosine are said to be *cofunctions;* * i.e., the cosine is the cofunction of the sine, and the sine is the cofunction of the cosine. Similarly, the tangent and cotangent are cofunctions, and the secant and cosecant are cofunctions.

---

*Not to be confused with *reciprocal* functions.

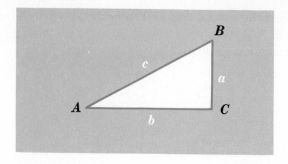

If A and B are the acute angles in right triangle ABC above, then they are *complementary* because their sum is 90°. Moreover,

$$\sin A = \frac{a}{c} = \cos B$$

$$\cos A = \frac{b}{c} = \sin B$$

Similarly, $\tan A = \cot B$ $\cot A = \tan B$
              $\sec A = \csc B$ $\csc A = \sec B$

Hence we have the following:

THEOREM   Any trigonometric function of an acute angle is equal to the cofunction of its complementary angle.

Thus $\sin 70° = \cos 20°$, $\cos 80° = \sin 10°$, $\tan 50° = \cot 40°$.

12 VARIATION OF THE FUNCTIONS OF AN ACUTE ANGLE

If r is fixed and if θ increases from 0° to 90° (Figure 11), then y increases and x decreases. It follows that *as an acute angle increases, its sine, tangent, and secant increase* while their cofunctions, the cosine, cotangent, and cosecant, decrease. Since neither leg of a right triangle can equal the hypotenuse, the sine and cosine of an acute angle must always be less than 1; the secant and cosecant must be greater than 1.

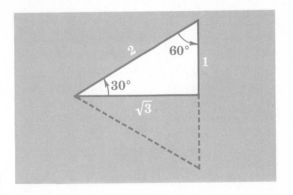

FIGURE 18

13 THE TRIGONOMETRIC FUNCTIONS OF 30°, 45°, 60°

Consider an equilateral triangle of side 2. The bisector of one of the 60° angles will also bisect the opposite side (Figure 18). By the pythagorean theorem, the length of the bisector is $\sqrt{3}$. Using Section 10, we find

$$\sin 30° = \frac{1}{2} \qquad\qquad \sin 60° = \frac{\sqrt{3}}{2}$$

$$\cos 30° = \frac{\sqrt{3}}{2} \qquad\qquad \cos 60° = \frac{1}{2}$$

$$\tan 30° = \frac{1}{\sqrt{3}} = \frac{\sqrt{3}}{3} \qquad \tan 60° = \sqrt{3}$$

The 30°-60°-90° triangle can be easily remembered if we note that the largest side (the hypotenuse) is twice the shortest side.

To compute the functions of 45°, draw an isosceles right triangle of leg 1 (Figure 19). The hypotenuse, by the pythagorean theorem, must be $\sqrt{2}$. Then, by Section 10,

$$\sin 45° = \frac{1}{\sqrt{2}} = \frac{\sqrt{2}}{2}$$

$$\cos 45° = \frac{1}{\sqrt{2}} = \frac{\sqrt{2}}{2}$$

$$\tan 45° = \frac{1}{1} = 1$$

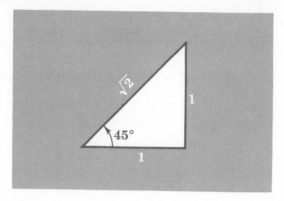

FIGURE 19

Because of their frequent occurrence, the following should be memorized, along with the figures from which they are derived:

$$\sin 30° = \cos 60° = \frac{1}{2}$$

$$\sin 45° = \cos 45° = \frac{\sqrt{2}}{2}$$

$$\sin 60° = \cos 30° = \frac{\sqrt{3}}{2}$$

As we shall see later, the tangent of any angle can be obtained by dividing its sine by its cosine. The three remaining functions can be obtained through their reciprocals.

The expression $\sin^2 \theta$ is a shorter way of writing $(\sin \theta)^2$. Since $\sin \theta$ is merely a number (the ratio of two lengths), we can speak of the square of this number and call it $\sin^2 \theta$. Thus

$$\sin^2 60° = \left(\frac{\sqrt{3}}{2}\right)^2 = \frac{3}{4} \qquad \sin^3 30° = \left(\frac{1}{2}\right)^3 = \frac{1}{8}$$

$$\sec^2 30° = \left(\frac{1}{\cos 30°}\right)^2 = \frac{1}{\cos^2 30°}$$

$$= \frac{1}{\left(\frac{\sqrt{3}}{2}\right)^2} = \frac{1}{\frac{3}{4}} = \frac{4}{3}$$

If we recall (Problem 15, Exercise 3) that $\sin 0° = 0$, $\cos 0° = 1$,

	0°	30°	45°	60°	90°
sin	$\dfrac{\sqrt{0}}{2}$	$\dfrac{\sqrt{1}}{2}$	$\dfrac{\sqrt{2}}{2}$	$\dfrac{\sqrt{3}}{2}$	$\dfrac{\sqrt{4}}{2}$
cos	$\dfrac{\sqrt{4}}{2}$	$\dfrac{\sqrt{3}}{2}$	$\dfrac{\sqrt{2}}{2}$	$\dfrac{\sqrt{1}}{2}$	$\dfrac{\sqrt{0}}{2}$

FIGURE 20

$\sin 90° = 1$, $\cos 90° = 0$, we can easily remember the sine and cosine of special first-quadrant angles by forming a mental picture of Figure 20.

EXERCISE 6   Do not use a calculator.

Compute the exact value of each of the following expressions. Leave the results in fractional form.

1. $\cos^5 60° + 6 \sin 45° \cos 45° - 3 \sin^3 90°$
2. $20 \sin^4 45° - 6 \cos^2 30° + 3 \sin^3 30° - \sin 0°$
3. $5\sqrt{3} \sin 60° - 7\sqrt{2} \cos 45° + \sin^4 30°$
4. $7 \cos^4 45° - \sin 60° \cos 30° + \cos 0°$
5. $12 \sin^2 60° - 16 \cos^4 30° + \sin 30°$
6. $3\sqrt{2} \cos 45° + 16 \sin^6 60° - 18 \cos 60°$
7. $3 \sin 45° \cos 30° + \sqrt{6} \cos^2 60° - \cos^3 90°$
8. $10 \cos^3 60° - 4 \sin^5 30° - \sin^2 45°$

Identify as true or false and give reasons.

9. $\cos 12° < \cos 34°$
10. $\sin 58° < \sin 61°$
11. $\sin 45° + \sin 45° = \sin 90°$
12. $\cot 20° = \tan 70°$
13. $\sin (5° + \theta) = \cos (85° - \theta)$
14. $\cos 50° = \dfrac{1}{\csc 50°}$
15. $\dfrac{1}{\sec 66°} = \sin 24°$
16. $\sec^2 12° = \dfrac{1}{\sin^2 78°}$

14 APPROXIMATIONS AND SIGNIFICANT FIGURES

If a given distance is *measured* and if its length is expressed in decimal form, it is conventional to write no more digits than are correct (or probably correct). Thus, if we say that the measured distance between points A and B is 17 meters, we mean that the result is given to the nearest meter; i.e., the true distance is closer to 17 meters than it is to 16 meters or 18 meters. This is an example of two-figure accuracy. If we say that the measured distance AB is 17.0 meters, we mean that the true distance is given to three significant figures, i.e., it is closer to 17.0 meters than it is to 16.9 or 17.1 meters. This implies that the true distance is somewhere between 16.95 and 17.05 meters. Notice that 17 and 17.0 do not mean the same thing when they represent approximate values.

The number of significant digits in a number is obtained by counting the digits from left to right, beginning with the first nonzero digit and ending with the rightmost digit. * Thus, 0.078060 has five significant digits, 70.00 has four, and 0.790 has only three. Notice that the number of significant digits does not depend on the position of the decimal point.

If calculations are made on approximate data, the result may have digits that are not significant. These digits should be rejected. The process of discarding these extra digits is called *rounding off* the number.

ILLUSTRATION 1   When 3.141592 is rounded off to five (significant) digits, we obtain 3.1416. In this case, 1 must be added to the fifth digit of the given number to obtain the best five-figure approximation. For four-figure and three-figure approximations, 3.141592 becomes 3.142 and 3.14, respectively.

ILLUSTRATION 2   In rounding off 78.25 to three-figure accuracy, we could logically get 78.2 or 78.3. In cases of this kind, it is conventional to *make the last digit an even number.* Thus, 78.25 rounds off to 78.2, and 78.15 rounds off to 78.2; however, 78.251 rounds off to 78.3.

---

*Ambiguity may result if the number in question is an integer ending in one or more 0's. For example, if the radius of the earth is given as 4000 mi, we may not know how many 0's are significant. If, however, the number 4000 was obtained from 3960 by rounding it off to the nearest multiple of 100 mi, then the first 0 is significant; the other two are not. In a case of this kind we usually use **scientific notation.** In this notation the number is expressed as a product. The first factor is the number formed by the significant digits with a decimal point placed after the first digit; the second factor is an integral power of 10. Thus, $4.0(10^3)$ indicates the number 4000, in which only the first 0 is significant. Also, $0.0123 = 1.23(10^{-2})$.

We shall adhere to the following two rules for rounding off the numbers that result when calculations are performed on approximate data.

1. When multiplication (or division) is performed on approximate data, *round off the result so that it will have as many significant figures as there are in the least accurate number in the data.*
2. When addition (or subtraction) is to be performed on approximate data, write one number beneath the other and add (or subtract); then round off the result so that the last digit retained is in the rightmost column in which both of the given numbers have significant digits.

If a small field is measured and found to be 11.3 meters long and 10.7 meters wide, we would be tempted to say that its area is $(11.3)(10.7) = 120.91$ square meters. To do so would be false accuracy. The result should be rounded off to three significant figures (the same as in the given data) to obtain 121 square meters. The first two figures in this result are correct, but the third is only a good approximation, because the true area is somewhere between $(11.25)(10.65) = 119.8125$ square meters and $(11.35)(10.75) = 122.0125$ square meters.

Since nearly all the angles listed in Table 1 (page 267) have trigonometric functions that are nonending decimals, the numbers appearing in this table are merely four-figure approximations. Hence most of the results obtained using this table will be approximations and should be considered as such. On a calculator there may be 8-, 10-, or 12-figure approximations, depending upon the type of calculator. However, if you are using a calculator to perform computations on three-figure data, then the results should be rounded off to only three-figure accuracy.

In solving triangles, we agree to set up the following correspondence between accuracy in sides and angles.

Accuracy in sides	Accuracy in angles
Two-figure	Nearest degree
Three-figure	Nearest tenth of a degree
Four-figure	Nearest hundredth of a degree

Hence within each of the following sets of data, the same degree of accuracy prevails:

23, 42, 62°
0.0461, 61.2°, 44.8°, 74.0°
8.624, 55.78°, 82.00°

If the data include a side with two-figure accuracy and another side with three-figure accuracy, then the computed parts should be written with only two-figure accuracy, which means that computed angles should be taken to the nearest degree. In general, *our results can be no more accurate than the least accurate item of the data.* If the given data include a number whose degree of accuracy is doubtful, we shall (in this book) *assume the maximum degree of accuracy.* For example, with no information to the contrary, the side 700 km (kilometers) will be treated as a three-figure number.

EXERCISE 7

Round off the following numbers and angles to (a) *four-figure accuracy,* (b) *three-figure accuracy, and* (c) *two-figure accuracy.*

1. 6.3725	**2.** 0.82948	**3.** 0.019074	**4.** 51.675
5. 21.2493°	**6.** 38.165°	**7.** 44.7350°	**8.** 95.4739°

The number 77.4 is the best three-figure approximation for all numbers from 77.35 to 77.45 inclusive. What range of numbers is covered by each of the following approximations?

9. 752	**10.** 2.49	**11.** 871	**12.** 0.0137
13. 0.9159	**14.** 4932	**15.** 6.400	**16.** 784.2

15 CALCULATOR OR TABLES?

To solve a triangle means to find from the given parts (sides and angles) the values of the remaining parts. Before we can solve a triangle, we need to consider the following two problems:

1. Given an angle, to find one of its trigonometric functions
2. Given a trigonometric function of some angle, to find the angle

Each of these two problems can be solved by using either a calculator or a table of trigonometric functions.

16 USING A CALCULATOR

EXAMPLE 1 Find sin 34° to four-figure accuracy.

SOLUTION   First, be certain that the degree-radian switch is in the "degree" position. Then press the following keys:

$$\boxed{3}\ \boxed{4}\ \boxed{\sin}$$

The displayed result is 0.5591929035. Hence, to four-figure accuracy,

$$\sin 34° = 0.5592$$

EXAMPLE 2 Find cot 72.1° to four-figure accuracy.

SOLUTION   Be sure the switch is set for degrees. Inasmuch as most calculators have no cotangent function, we observe that $\cot 72.1° = \dfrac{1}{\tan 72.1°}$. Press the following keys:

$$\boxed{7}\ \boxed{2}\ \boxed{\cdot}\ \boxed{1}\ \boxed{\tan}\ \boxed{1/x}\ ^*$$

An alternative procedure (leading to the same final result) is to press

$$\boxed{1}\ \boxed{\div}\ \boxed{7}\ \boxed{2}\ \boxed{\cdot}\ \boxed{1}\ \boxed{\tan}\ \boxed{=}$$

The result 0.3229911993 is now displayed. Therefore, to four-figure accuracy,

$$\cot 72.1° = 0.3230$$

EXAMPLE 3 If cos θ = 0.3616, find θ to four-figure accuracy.

SOLUTION   If u is a positive number between 0 and 1, then Arccos u or Cos^{-1} u (read "inverse cosine u") designates the first-

---

*This reciprocal key finds the reciprocal of the displayed number, in this case, tan 72.1°.

quadrant angle whose cosine is u. Thus, Arccos $\frac{1}{2} = 60°$ and Cos$^{-1} \frac{\sqrt{2}}{2} = 45°$. If cos $\theta = 0.3616$, then $\theta = $ Cos$^{-1} 0.3616$. With the switch set for degrees, press these keys:

$$\boxed{\cdot}\,\boxed{3}\,\boxed{6}\,\boxed{1}\,\boxed{6}\,\boxed{\text{arc or inv}}^{*}\,\boxed{\text{cos}}^{*}$$

The displayed result is 68.80150994. Thus, to four-figure accuracy,

$$\theta = 68.80°$$

EXAMPLE 4 If cot $\theta = 1.511$, find θ to four-figure accuracy.

SOLUTION Since the cotangent function does not appear on calculators, we recall that if cot $\theta = 1.511$, then tan $\theta = \dfrac{1}{1.511}$. Accordingly, with the switch set for degrees, we press

$$\boxed{1}\,\boxed{\cdot}\,\boxed{5}\,\boxed{1}\,\boxed{1}\,\boxed{1/x}\,\boxed{\text{inv}}\,\boxed{\text{tan}}$$

The displayed result is 33.49712364, and to four-figure accuracy,

$$\theta = 33.50°$$

17 A TABLE OF TRIGONOMETRIC FUNCTIONS

In Table 1 (in the table section at the end of the book) there are listed, to four decimal places (for numbers less than 1) or four significant figures (for numbers greater than 1), the sine, cosine, tangent, and cotangent for acute angles at intervals of one-tenth of a degree. For angles less than 45°, find the name of the function at the *top* of the column, then read *down* until the angle is found at the *left*. For angles greater than 45°, find the name of the function at the *bottom* of the column, then read *up* until the angle is found at the *right*.

---

*On some calculators only one key is needed: $\boxed{\text{Cos}^{-1}}$.

Remember that results obtained using a four-place table will not have more than four-figure accuracy no matter how high the degree of accuracy of the given data.

18 GIVEN AN ANGLE, TO FIND ONE OF ITS FUNCTIONS

EXAMPLE 1 Find cos 82.6°.

SOLUTION In Table 1, in the column with *cos* at its *foot,* move up to the number in line with 82.6°. Thus cos 82.6° = 0.1288.

EXAMPLE 2 Find tan 27.3°.

SOLUTION In Table 1, in the column with *tan* at its *head,* come down to the number in line with 27.3°. Hence tan 27.3° = 0.5161.

19 GIVEN A FUNCTION OF AN ANGLE, TO FIND THE ANGLE

EXAMPLE 1 Find θ if sin θ = 0.9385.

SOLUTION Since sines are found in column 3 reading *down* and in column 6 reading *up,* we must search through these two columns for the number 0.9385. It appears in the sixth column, which has *sin* at its *foot.* This column contains the sines of the angles in the *right* column. On a line with 0.9385, we find in the *right* column the angle 69.8°. Hence,

if $\qquad\qquad$ sin θ = 0.9385

then $\qquad\qquad$ θ = 69.8°

EXAMPLE 2 Find θ if cot θ = 1.638.

SOLUTION We search the two cotangent columns, the fourth going up and the fifth going down, and find 1.638 in the fifth column. Since this column has *cot* at its *head,* we associate 1.638 with the angle at the *left.* Hence,

if $\qquad\qquad$ cot θ = 1.638

then $\qquad\qquad$ θ = 31.4°

The student should guard against writing $\cot \theta = 1.638 = 31.4°$. The second equality sign is incorrectly used because 1.638 does *not* equal 31.4° and $\cot \theta$ does *not* equal 31.4°.

EXERCISE 8

Use a calculator or a four-place table (Table 1) to find the value of each of the following to four decimal places if the number is less than 1 and to four significant digits if the number is greater than or equal to 1.

1. $\tan 29.7°$	**2.** $\cos 9.1°$	**3.** $\cot 83.4°$
4. $\sin 74.8°$	**5.** $\cos 53.2°$	**6.** $\sin 34.0°$
7. $\tan 66.9°$	**8.** $\cot 7.5°$	**9.** $\cot 18.3°$
10. $\tan 44.6°$	**11.** $\sin 49.8°$	**12.** $\cos 52.1°$

Use a calculator or a four-place table (Table 1) to find θ, to the nearest tenth of a degree, from each of the following functions of θ.

13. $\sin \theta = 0.7145$	**14.** $\cot \theta = 0.0629$
15. $\cos \theta = 0.9444$	**16.** $\tan \theta = 0.5520$
17. $\cos \theta = 0.9976$	**18.** $\sin \theta = 0.5402$
19. $\tan \theta = 4.511$	**20.** $\cot \theta = 0.5475$
21. $\tan \theta = 3.096$	**22.** $\cos \theta = 0.6320$
23. $\cot \theta = 2.488$	**24.** $\sin \theta = 0.6909$

20 INTERPOLATION

When a sports announcer says, "The ball is on the 27-yd line," most football fans realize that the announcer estimates that the ball is $\frac{2}{5}$ of the way from the 25-yd line to the 30-yd line. This process of literally "reading between the lines" is called interpolation. Another example is, "Interpolate to approximate the value of $\sqrt{8}$." Knowing $\sqrt{4} = 2$ and $\sqrt{9} = 3$, we conclude that $\sqrt{8}$ is a number between 2 and 3. Moreover, 8 is $\frac{4}{5}$ of the way from 4 to 9. Assume that for a small increase in a number N, the change in $\sqrt{N}$ is proportional to the change in N. Then $\sqrt{8}$ would lie $\frac{4}{5}$ of the way from 2 to 3. Since $\frac{4}{5}$ of 1 is 0.8, we conclude that $\sqrt{8}$ is approximately 2.8. This result is correct to only one decimal place. The process of interpolation is important in all work involving the use of tables.

We already know that the trigonometric functions do not change uniformly with the change in the angle (if an angle is doubled, its sine does not double). But if the angle is changed by only a few hundredths of a degree, the change in the function is very nearly proportional to the change in the angle.

EXAMPLE 1 Find $\sin 65.84°$.

SOLUTION Here we must interpolate between $65.80°$ and $65.90°$.

$$\begin{array}{lll} \sin 65.80° & = 0.9121 \\ \sin 65.84° & = \\ \sin 65.90° & = 0.9128 \end{array}$$

As the angle increases $0.10°$ (from $65.80°$ to $65.90°$), its sine increases 7 ten-thousandths. Our angle is $\frac{4}{10}$ of the way from $65.80°$ to $65.90°$. Hence the sine of our angle should be $\frac{4}{10}$ of the way from 0.9121 to 0.9128. But $\frac{4}{10}(7) = 2\frac{4}{5} \rightarrow 3$. (Round off to 3 because $2\frac{4}{5}$ is closer to 3 than it is to 2.) Since the sine is increasing, *add* the 3 to 0.9121 to get

$$\sin 65.84° = 0.9124$$

EXAMPLE 2 Find $\cos 15.17°$.

SOLUTION

$$\begin{array}{lll} \cos 15.10° & = 0.9655 \\ \cos 15.17° & = \\ \cos 15.20° & = 0.9650 \end{array}$$

An *increase* of $0.10°$ in the angle produces a *decrease* of 5 ten-thousandths in the cosine. Our angle is $\frac{7}{10}$ of the way from $15.10°$ to $15.20°$. Hence we want $\frac{7}{10}$ of the decrease of 5. But $\frac{7}{10}(5) = 3\frac{1}{2} \rightarrow 3$. * *Subtracting* this number from 0.9655

---

*Since $3\frac{1}{2}$ is exactly halfway between 3 and 4, we choose the number that makes the last digit of the *final result* an even number. (See Illustration 2, Section 14.) Hence we choose 0.9652 rather than 0.9651.

gives

$$\cos 15.17° = 0.9652$$

EXAMPLE 3 Find θ if $\tan \theta = 0.6206$.

SOLUTION

$$\tan 31.80° = 0.6200$$
$$\tan \theta = 0.6206$$
$$\tan 31.90° = 0.6224$$

Our number 0.6206 is $\frac{6}{24}$ of the way from 0.6200 to 0.6224. Hence θ should be $\frac{6}{24}$ of the way from 31.80° to 31.90°. But $\frac{6}{24}(10) = 2\frac{1}{2} \to 2.$* Hence

$$\theta = 31.82°$$

EXAMPLE 4 Find θ if $\cos \theta = 0.2810$.

SOLUTION

$$\cos 73.60° = 0.2823$$
$$\cos \theta = 0.2810$$
$$\cos 73.70° = 0.2807$$

Our number 0.2810 is $\frac{13}{16}$ of "the way down." Hence θ is $\frac{13}{16}(10) = 8\frac{1}{8} \to 8$ away from 73.60°. Therefore

$$\theta = 73.68°$$

It is to be noted that, in this three-line method of interpolation, *the small angle is always written on top. All the differences are measured from the small angle and its function.*

EXERCISE 9

(This exercise is to be used only with Table 1. It is unnecessary if you are using a calculator.)

*Number is exactly halfway. Make last digit even.

Use a four-place table (Table 1). Interpolate to find the value of each of the following.

1. $\cos 48.23°$
2. $\sin 12.48°$
3. $\cot 55.94°$
4. $\tan 21.07°$
5. $\cot 22.68°$
6. $\tan 76.39°$
7. $\cos 38.52°$
8. $\sin 69.84°$
9. $\sin 29.97°$
10. $\cot 57.75°$
11. $\tan 34.13°$
12. $\cos 65.11°$
13. $\tan 81.72°$
14. $\cos 23.84°$
15. $\sin 72.36°$
16. $\cot 13.09°$

Interpolate to find θ to the nearest hundredth of a degree.

17. $\tan \theta = 0.3029$
18. $\cos \theta = 0.0240$
19. $\sin \theta = 0.6058$
20. $\cot \theta = 0.4571$
21. $\cos \theta = 0.7383$
22. $\sin \theta = 0.9828$
23. $\cot \theta = 6.538$
24. $\tan \theta = 1.593$
25. $\cot \theta = 0.3318$
26. $\tan \theta = 0.8886$
27. $\cos \theta = 0.4806$
28. $\sin \theta = 0.3742$
29. $\sin \theta = 0.8997$
30. $\cot \theta = 1.932$
31. $\tan \theta = 3.429$
32. $\cos \theta = 0.9804$

21 THE SOLUTION OF RIGHT TRIANGLES

To solve a triangle means to find from the given parts the values of the remaining parts. A right triangle is determined by

1. Two of its sides, or
2. One side and an acute angle

In either case it is possible to find the remaining parts by using the special definitions in Section 10 together with the fact that the acute angles of a right triangle are complementary. For convenience we list here again these special definitions.

For any acute angle θ lying in a right triangle:

$$\sin \theta = \frac{\text{opp}}{\text{hyp}}$$

$$\cos \theta = \frac{\text{adj}}{\text{hyp}}$$

$$\tan \theta = \frac{\text{opp}}{\text{adj}}$$

For any triangle we shall use the small letters a, b, and c to denote the lengths of the sides that are opposite the angles A, B, and C, respectively. In a right triangle we shall always reserve the letter c for the hypotenuse. Hence $C = 90°$.

EXAMPLE 1 Solve the right triangle having an acute angle of 38.7°, the side adjacent to the angle being 311.

SOLUTION PLAN We first draw the triangle to scale and label numerically the parts that are known (Figure 21). Then

(1) $$B = 90° - 38.7°$$
$$= 51.3°$$

(2) To find a, we observe that the given side and the required side are related to the given angle by the equation

$$\tan 38.7° = \frac{a}{311}$$

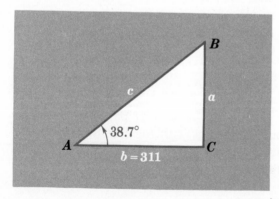

FIGURE 21

Multiply both sides of the equation by 311:

$$311 \tan 38.7° = a$$

Hence
$$a = 311 \tan 38.7°$$

(3) To find c, we notice that the given parts, 38.7° and 311, are related to the required part through the cosine of the angle:

$$\cos 38.7° = \frac{311}{c}$$

Multiply both sides by c:

$$c \cos 38.7° = 311$$

Divide both sides by $\cos 38.7°$:

$$c = \frac{311}{\cos 38.7°}$$

CALCULATOR SOLUTION

(2) $a = 311 \tan 38.7°$

Using an algebraic operating system (AOS) with the switch set for degrees, we press

$$\boxed{3}\ \boxed{1}\ \boxed{1}\ \boxed{\times}\ \boxed{3}\ \boxed{8}\ \boxed{\cdot}\ \boxed{7}\ \boxed{\tan}\ \boxed{=}$$

Using Reverse Polish Notation (RPN) with the switch set for degrees, we press

$$\boxed{3}\ \boxed{1}\ \boxed{1}\ \boxed{\text{ENTER}}\ \boxed{3}\ \boxed{8}\ \boxed{\cdot}\ \boxed{7}\ \boxed{\tan}\ \boxed{\times}$$

In each case the displayed result is 249.1579829. Rounding off to three figures, we get

$$a = 249$$

(3)
$$c = \frac{311}{\cos 38.7°}$$

We press:

Algebraic logic $\boxed{311}$ $\boxed{\div}$ $\boxed{38.7}$ $\boxed{\cos}$ $\boxed{=}$

RPN $\boxed{311}$ $\boxed{\text{ENTER}}$ $\boxed{38.7}$ $\boxed{\cos}$ $\boxed{\div}$

In both cases the displayed result is 398.4980558, which, rounded off to three significant figures, gives

$$c = 398$$

TABLE SOLUTION

(2) $a = 311 \tan 38.7°$
    $= 311(0.8012)$ from Table 1
    $= 249.1732 \rightarrow 249$

(3) $c = \dfrac{311}{\cos 38.7°} = \dfrac{311}{0.7804}$ from Table 1
    $= 398.514 \rightarrow 399$

If you have a calculator that does not deal with trigonometric functions but does multiply and divide, it can be used to compute a and c after tan 38.7° and cos 38.7° have been obtained from the table. Observe that the calculator answer for c is different from the table result. The calculator result, of course, is more accurate.

This problem illustrates three-figure accuracy in the data and the computed results.

EXAMPLE 2 Solve the right triangle whose hypotenuse is 40.50 and one of whose legs is 34.56.

SOLUTION PLAN Draw the triangle (Figure 22) and label numerically the given parts.

(1) Since the hypotenuse and the side opposite A are given,

$$\sin A = \frac{34.56}{40.50}$$

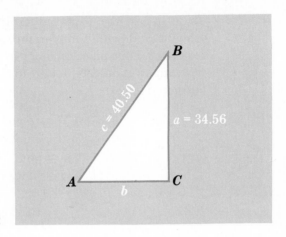

FIGURE 22

or
$$A = \text{Sin}^{-1}\frac{34.56}{40.50}$$

(2)
$$B = 90° - A$$

(3) To find b, use

$$\cos A = \frac{b}{40.50}$$

which is equivalent to

$$b = 40.50 \cos A$$

CALCULATOR SOLUTION

(1)
$$A = \text{Sin}^{-1}\frac{34.56}{40.50}$$

Using an algebraic logic calculator with the switch set for degrees, we press

$$\boxed{34.56}\ \boxed{\div}\ \boxed{40.50}\ \boxed{=}\ \boxed{\text{inv}}\ \boxed{\sin}$$

Using RPN with the switch set for degrees, we press

$$\boxed{34.56}\ \boxed{\text{ENTER}}\ \boxed{40.50}\ \boxed{\div}\ \boxed{\text{inv}}\ \boxed{\sin}$$

In each case the displayed result is 58.57609351, which we should jot down, or store in the calculator, for further use in finding b. Four-place accuracy in the given sides implies that the angles should be found to the nearest hundredth of a degree. Hence

$$A = 58.58°$$

(2)
$$B = 90° - 58.58°$$
$$= 31.42°$$

(3)
$$b = 40.50 \cos A$$

Assuming that the "unrounded off" value of A is still displayed on the calculator, we press

Algebraic logic $\boxed{\cos}$ $\boxed{\times}$ $\boxed{40.50}$ $\boxed{=}$

RPN $\boxed{\cos}$ $\boxed{\text{ENTER}}$ $\boxed{40.50}$ $\boxed{\times}$

The displayed result is 21.11531198. Hence, to four-place accuracy,

$$b = 21.12$$

Check Since $a^2 + b^2 = c^2$, then $b = \sqrt{c^2 - a^2}$. If the calculator has a *square key* and a *square root key*, for an algebraic logic instrument we press

$\boxed{40.50}$ $\boxed{x^2}$ $\boxed{-}$ $\boxed{34.56}$ $\boxed{x^2}$ $\boxed{=}$ $\boxed{\sqrt{x}}$

The result on display is 21.11531198, which agrees with our value for b.

TABLE SOLUTION

(1)
$$\sin A = \frac{34.56}{40.50} = 0.8533$$

This result may be obtained by actual division or by using a calculator that merely performs the four basic algebraic operations. Since the given sides have four-place accuracy, angle A should be found to the nearest hundredth of a degree. Using Table 1 and interpolating, we find

$$A = 58.57°$$

(2)
$$B = 90° - 58.57°$$
$$= 31.43°$$

(3)
$$b = 40.50 \cos 58.57°$$
$$= 40.50(0.5214) \qquad \text{from Table 1}$$
$$= 21.1167 \to 21.12$$

Check   The value of b could have been found by using

$$\tan A = \frac{34.56}{b}$$

$$b = \frac{34.56}{\tan 58.57°} = \frac{34.56}{1.636} = 21.1247 \to 21.12$$

It may be easier to use

$$b = \frac{34.56}{\tan 58.57°} = 34.56 \cot 58.57° = 34.56(0.6111)$$

$$= 21.1196 \to 21.12$$

Observe that the calculator and table solutions disagree on the value of A.

This problem illustrates four-figure accuracy in the data and the results.

EXERCISE 10

Use a calculator or a four-place table (Table 1) to solve the following right triangles. The answers at the back of the book were obtained using a calculator. If a table is used, the results may vary slightly.

1. $A = 56.1°$, $a = 166$
2. $B = 21.1°$, $b = 32.4$
3. $B = 42°$, $a = 3.7$
4. $A = 53°$, $b = 48$
5. $B = 39.80°$, $c = 33.22$
6. $A = 72.6°$, $c = 610$

7. $a = 537, b = 202$

8. $a = 4.190, b = 8.750$

9. $b = 40, c = 61$

10. $a = 1906, c = 1980$

11. $b = 247.6, c = 561.0$

12. $a = 0.592, c = 0.730$

13. $a = 2400, b = 5520$

14. $a = 72, b = 31$

15. $A = 82°, c = 99$

16. $B = 38.71°, c = 5051$

17. $a = 2.53, c = 8.00$

18. $A = 67.80°, a = 0.9020$

19. $B = 17.42°, b = 2345$

20. $a = 246, b = 135$

22 ANGLES OF ELEVATION AND DEPRESSION; BEARING OF A LINE

The *angle of* $\left\{ \begin{matrix} elevation \\ depression \end{matrix} \right\}$ of a point P as seen by an observer O is the

vertical angle measured from the horizontal line through O $\left\{ \begin{matrix} upward \\ downward \end{matrix} \right\}$

to the line of sight OP (see Figure 23).

In surveying and some kinds of navigation, the *bearing* of a line in a horizontal plane is the *acute* angle made by this line with a north-south line. In giving the bearing of a line, write first the letter N or S, then the angle of deviation from north or south, then the letter E or W. Thus, in Figure 24 the bearing of line OA is N 70° E, or the bearing of point A from point O is N 70° E.

EXAMPLE 1 A vertical stake 20.0 cm high casts a horizontal shadow 12.5 cm long. What time is it if the sun rose at 6:00 A.M. and will be directly overhead at noon?

SOLUTION The angle of elevation of the sun (Figure 25) is found by

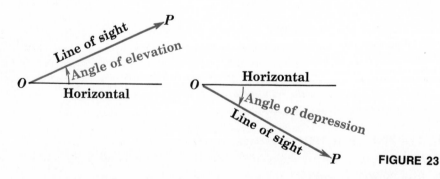

FIGURE 23

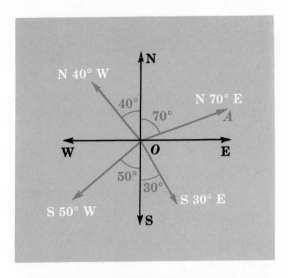

FIGURE 24

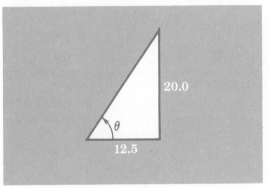

FIGURE 25

$$\tan \theta = \frac{20.0}{12.5} = 1.600$$

$$\theta = 58.0°$$

It takes the earth 6 h to rotate through 90°. Since this rotation is uniform, each degree of elevation of the sun will correspond to $\frac{6}{90}$ of an hour, or 4 min. Consequently a rotation through 58.0° will require (58)(4 min) = 232 min = 3 h and 52 min. Hence the time is 9:52 A.M.

EXAMPLE 2 From a lookout tower A a column of smoke is sighted due south. From a second tower B, 5.00 mi west of A, the smoke

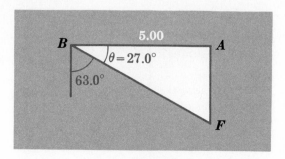

FIGURE 26

is observed in the direction S 63.0° E. How far is the fire from B? From A? (See Figure 26.)

SOLUTION   Angle $FBA = \theta = 90° - 63.0° = 27.0°$. To get BF, use

$$\cos 27.0° = \frac{5.00}{BF}$$

Hence $$BF = \frac{5.00}{\cos 27.0°} = \frac{5.00}{0.8910}$$

$$= 5.61 \text{ mi}$$

To obtain AF, use

$$\tan 27.0° = \frac{AF}{5.00}$$

Hence $AF = 5.00 \tan 27.0° = (5.00)(0.5095)$
$$= 2.55 \text{ mi}$$

EXERCISE 11

1. From a point on the ground 120 meters away from the foot of the Eiffel Tower, the angle of elevation of the top of the tower is 68.2°. How high is the tower?

2. From a point 701 ft from, and in the same horizontal plane with, the top of Bridalveil Falls in Yosemite National Park, the angle of depression of the foot of the falls is 41.5°. Find the height of Bridalveil Falls.

3. A helicopter hovers 400 ft above one end of a bridge that spans the Missouri River at Jefferson City, Mo. The angle of depression of the other end of the bridge from the helicopter is 32.0°. How long is the bridge?

4. The angle of elevation of the top of City Hall Tower in Philadelphia from a point on level ground 300 ft from its base is 61.3°. How high is the building?

5. Cincinnati, Ohio, is due south of East Lansing, Mich. Racine, Wis., is due west of East Lansing and is 300 mi N 34° W from Cincinnati. How far is East Lansing from Cincinnati? From Racine? *

6. Gainesville, Fla., is 110 mi due south of Waycross, Ga. Dothan, Ala., is 180 mi due west of Waycross. What is the bearing of Dothan from Gainesville? Gainesville from Dothan? How far is Gainesville from Dothan? *

7. San Bernardino, Calif., is 100 mi due north of San Diego. Yuma, Ariz., is S 56° E from San Bernardino and due east of San Diego. How far is Yuma from San Bernardino? From San Diego? *

8. Greenwood, Miss., is 110 mi due south of West Memphis, Ark. Fayetteville, Tenn., is due east of West Memphis and 230 mi from Greenwood. What is the bearing of Fayetteville from Greenwood? Greenwood from Fayetteville? How far is West Memphis from Fayetteville? *

9. From a cottage C on the top of a hill that is 120 meters above sea level, the angle of depression of a sailboat S is 31.0°. Find the length of the airline distance CS from the cottage to the sailboat.

10. Each of the equal sides of an isosceles triangle is 70.00 meters. Each of the equal angles is 35.00°. Find the altitude and the base of the triangle.

11. A flagpole stands atop a 65-ft building. From the position of an observer whose eyes are 5 ft above the ground, the angles of elevation of the top and bottom of the flagpole are 47° and 41°, respectively. Find the length of the flagpole.

12. The angle of elevation of a ladder leaning against a wall is 82°. Find the length of the ladder if its foot is 2.5 ft from the wall.

---

*Ignore the curvature of the earth and assume only two-place accuracy. Get angles to the nearest degree and distances to the nearest multiple of 10 mi.

13. The angle made by a taut kite string with the horizontal is 35°. How high is the kite when 140 ft of string have been paid out?

14. Let θ be the acute angle of intersection of two lines, l and m. Let AB be any segment of line l. If AC and BD are perpendiculars drawn to line m, then CD is called the projection of AB on line m. Show that $CD = AB \cos \theta$. Investigate this relationship for values of θ close to 0°; for θ close to 90°.

15. Find the radius of a circle in which a chord of 10.8 cm subtends a central angle of 38.4°.

16. A railroad track makes an angle of 6.3° with the horizontal. How many meters does a locomotive rise while traveling 100 meters along the track?

17. Find the perimeter of a regular polygon of 30 sides inscribed in a circle of radius 1.000 meter. Compare this number with the circumference of the circle.

18. A surveyor, running a line due south, discovered that she had to change her course to bypass a thicket. From point A she measured 200.0 meters in the direction S 31.00° E to point B. (*a*) How many meters should she measure from B in the direction S 51.00° W to arrive at a point C which is due south of A? (*b*) Find the distance from A to C.

19. A vertical stake 77.0 cm high casts a horizontal shadow 50.0 cm long. What time is it if the sun was directly overhead at 12:30 P.M. and will set at 6:30 P.M.?

20. A flagpole 9.10 meters high casts a horizontal shadow 25.0 meters long. What time is it if the sun rose at 5:45 A.M. and will be directly overhead at 11:45 A.M.?

21. Find, to the nearest degree, the angle made by a diagonal of a cube with an edge that meets it.

22. A motorboat moving in the direction S 40° W passes a buoy that is due east of a television tower. If the angle of elevation of the top of the tower from the buoy is 23°, what is the angle of elevation when the ship is closest to the tower?

23. A board fence of height h runs east and west. The angle of elevation of the sun is θ and its bearing is S ϕ W. Find the width of the shadow cast by the fence on level ground.

24. From the top of a cliff at the edge of a lake, the angle of depression of a nearby buoy is θ and that of a second buoy directly behind is ϕ. If the closer buoy is a meters from the base of the cliff, show that the

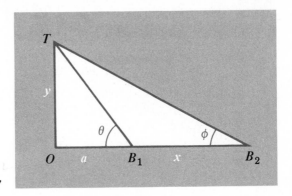

FIGURE 27

distance between the buoys is $a(\tan \theta \cot \phi - 1)$ meters. *Hint:* See Figure 27. Show that $y = a \tan \theta$, $a + x = y \cot \phi$. Then eliminate y from this system of equations and solve for x.

25. A helicopter hovers directly above an east-west road that is on level ground. Looking east, the pilot sees, with an angle of depression θ, a pothole in the road. Looking west, he observes, with an angle of depression ϕ, another pothole in the road. If the potholes are a meters apart, find the height of the helicopter.

26. An observer at A looks due north and sees a meteor with an angle of elevation of $55°$. At the same instant, another observer, 10 km west of A, sees the same meteor and approximates its position as N $50°$ E but fails to note its angle of elevation. Find the height of the meteor and its distance from A.

23 VECTORS

A *vector quantity* is a quantity having magnitude and direction. Examples of vector quantities are forces, velocities, accelerations, and displacements. A *vector* is a directed line segment. A vector quantity can be represented by means of a vector if (1) the direction of the vector is the same as that of the vector quantity and (2) the length of the vector represents, to some convenient scale, the magnitude of the vector quantity. For example, a velocity of 30 mi/h in a northerly direction can be represented by a 3-in line segment pointed north.

The *resultant* (or vector sum) of two vectors is the diagonal of a parallelogram having the two given vectors as adjacent sides. In physics it is shown that like vector quantities, such as forces, are combined according to this vector law of addition. In Figure 28, vector OR is the resultant

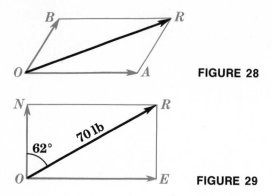

FIGURE 28

FIGURE 29

of vectors OA and OB. If the lengths and directions of OA and OB are known, then the length and direction of OR can be found by solving triangle OAR.

Two vectors that have a certain resultant are said to be *components* of that resultant. If the two vectors are at right angles to each other, they are called *rectangular components*. It is possible to resolve a given vector into components along any two different specified directions. For example (Figure 29), a force of 70 lb acting in the direction N 62° E can be resolved into an easterly force and a northerly force by solving the right triangle ORN.

In air navigation, the bearing of a line OC is the angle measured *clockwise* from the north to the line OC. Referring to Figure 30, in aviation the bearing of line OC is 110°, or the direction OC is 110°; the bearing of OD is 260°.

EXAMPLE 1 An airplane with an *air speed* (speed in still air) of 178 mi/h is headed due south. (Its *heading* is 180.0°.) If a west wind of 27.5 mi/h is blowing, find the *course* of the plane (the direc-

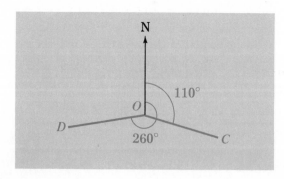

FIGURE 30

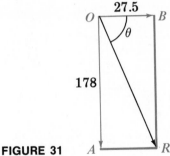

FIGURE 31

tion it travels) and its *ground speed* (actual speed with respect to the ground).

SOLUTION   To find θ (see Figure 31), use

$$\tan \theta = \frac{178}{27.5}$$

or $\theta = 81.22° \rightarrow 81.2°$

To find OR, use

$$\cos \theta = \frac{27.5}{OR}$$

Hence $OR = \dfrac{27.5}{\cos \theta} = 180.1 \rightarrow 180$

Inasmuch as the original data indicate only three-figure accuracy, the results should be rounded off to three significant digits. Hence the course of the plane is $90.0° + 81.2° = 171.2°$, and its ground speed is 180 mi/h.

EXAMPLE 2   What is the minimum force required to prevent a 90-lb barrel from rolling down a plane that makes an angle of 17° with the horizontal? Find the force of the barrel against the plane.

SOLUTION   The weight or force of 90 lb, OR, acting vertically downward can be resolved into two rectangular components, one

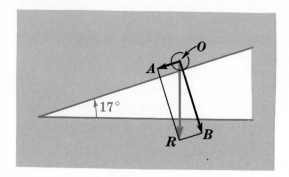

FIGURE 32

parallel to the plane and the other perpendicular to it (see Figure 32). Hence

$$OA = OR \sin \angle ORA * = 90 \sin 17° = 26.3$$

The force required to prevent the barrel from rolling is 27 lb (to the nearest pound).

The force of the barrel against the plane is

$$OB = OR \cos \angle ROB = 90 \cos 17° = 86 \text{ lb}$$

EXERCISE 12

1. Morgantown, W.Va., is 57 mi due south of Pittsburgh, Pa. A helicopter leaves Pittsburgh at 1:00 P.M. and travels at 70 mi/h with a bearing of 160°. When (to the nearest minute) will it be closest to Morgantown?

2. Nazareth, Pa., is 60 mi due west of Newark, N.J. An airplane leaves Newark at 10:00 A.M. and flies 100 mi/h with a heading of 295°. When (to the nearest minute) will the plane be due north of Nazareth?

3. An airplane leaves an airport at noon and flies 240 km/h on a bearing of 55°. A second plane leaves the same airport at 1:00 P.M. and flies 320 km/h on a bearing of 145°. Find the bearing of the first plane from the second one at 2:00 P.M.

---

* $\angle ORA = 17°$ because they are acute angles with their sides respectively perpendicular.

4. A plane leaves point A and travels 33 km on a bearing of 110°; it then turns to a heading of 200°. If the plane stops when it is due south of A, how far is it from the starting point?

5. The wind is blowing from the north at 54.5 km/h. The pilot of an airplane finds that if she heads her plane in the direction 74.0°, she will move due east. Find the plane's air speed and ground speed.

6. An airplane with an air speed of 160 km/h is headed due west. Because of a south wind, the plane's ground speed is 170 km/h. Find the plane's course (the direction it actually moves) and the wind speed.

7. A west wind is blowing. If a helicopter has a heading of 234.0°, it will move due south at 20.0 mi/h. Find the wind speed and the air speed of the helicopter.

8. An east wind causes a small plane, flying with an air speed of 70 mi/h, to travel due south with a ground speed of 60 mi/h. Find the plane's heading and the wind speed.

9. A wind of 30.0 mi/h is blowing in the direction 215.0°. A pilot wants his plane to move in the direction 305.0° with a ground speed of 200 mi/h. Find the proper heading and air speed.

10. The cruising speed of a boat with an outboard motor is 660 ft/min in still water. If the boat is headed S 70.0° W, it will move due west across a river that flows due north. Find the time required for the boat to cross the river if the river is 1861 ft wide.

11. A motorboat capable of a speed of 600 meters/min in still water is at the south bank of a river that flows west at 132 meters/min. (*a*) In what direction should the boat be steered if it is to move due north across the river? (*b*) With what speed does the boat cross the river?

12. A tanker moves due east at 16.5 mi/h. A jogger runs across its deck from north to south at 7.75 mi/h. Find the runner's direction and speed relative to the earth's surface.

13. A bus weighing 2700 kg stands on a hillside. If a force of 446 kg is required to keep the bus from rolling down, what angle does the hillside make with the horizontal?

14. A barrel resting on an inclined plane exerts a force of 200 kg against the plane. A force of 138 kg is needed to prevent the barrel from rolling down. (*a*) What angle does the plane make with the horizontal? (*b*) Find the weight of the barrel.

15. A truck weighing 5530 lb travels up a bridge that makes an angle of 11.70° with the horizontal. (*a*) Find the force exerted by the truck

against the bridge. (*b*) If the truck stops, what force is necessary to prevent it from rolling back downhill?

16. An inclined ramp makes an angle of 10.2° with the horizontal. A car on the ramp exerts a force of 1800 lb against the ramp. (*a*) Find the weight of the car. (*b*) What force is needed to keep the car from moving down the ramp?

17. A child is pushing a toy sailboat with a stick that makes an angle of 19° with the horizontal. (*a*) If the push along the stick is 2.3 kg, what force tends to move the sailboat? (*b*) What force tends to submerge it?

18. A car traveling 55 mi/h encounters rain that is falling straight down. Find the vertical speed of the raindrops if the streaks they form on the side windows make an angle of 62° with the vertical.

19. Two people, walking along a straight path, are carrying a heavy object between them. One person exerts a force of 60 lb at an angle of 10° with the vertical. (*a*) Find the force exerted by the other person pushing at an angle of 15° with the vertical. (*b*) Find the weight of the object. *Hint:* The horizontal components of the two forces must counterbalance each other.

20. Three forces of 11.11 kg, 55.55 kg, and 88.88 kg act in the directions S 20.20° W, S 34.34° W, and S 67.67° W, respectively. Find the magnitudes of two forces, one acting due north and the other acting due east, which will counterbalance these three forces. *Hint:* The force that acts due north must neutralize the southerly components of the three given forces.

3 TRIGONOMETRIC IDENTITIES

24 THE FUNDAMENTAL RELATIONS

In Section 9 we discussed the problem of determining all the trigonometric functions of an angle if one of them is given. This was a geometric process involving the construction of a triangle of reference for the angle. We shall now consider purely analytic relations among the functions themselves. These relations are of considerable importance in other branches of mathematics as well as in engineering and physics. Very often, the key to the solution of some important problem will be your ability to replace a given mathematical expression with an equivalent one that is more readily usable. Many such problems will involve the trigonometric functions. Consequently, the basic transformations which you will learn to use in this chapter are among the most important things that you should carry with you from this course into more advanced work in mathematics and science.

For any* angle θ, the following eight fundamental relations are true:

$$\csc \theta = \frac{1}{\sin \theta} \tag{1}$$

$$\sec \theta = \frac{1}{\cos \theta} \tag{2}$$

---

*Strictly speaking, for every angle for which the functions actually exist. For example, (4) has no meaning when $\theta = 90°$ because $\cos 90° = 0$; hence $\tan 90° = \dfrac{\sin 90°}{\cos 90°} = \dfrac{1}{0}$, which does not exist. Likewise (8) has no meaning for $\theta = 180°$ because $\cot 180°$ and $\csc 180°$ do not exist (see Section 8). Exceptions like these *can* occur only for $\theta = 0°$, $90°$, $180°$, $270°$, and angles coterminal with them. Notice, however, that (6) does hold for $\theta = 180°$: $\sin^2 180° + \cos^2 180° = 0^2 + (-1)^2 = 1$.

$$\cot \theta = \frac{1}{\tan \theta} \tag{3}$$

$$\tan \theta = \frac{\sin \theta}{\cos \theta} \tag{4}$$

$$\cot \theta = \frac{\cos \theta}{\sin \theta} \tag{5}$$

$$\sin^2 \theta + \cos^2 \theta = 1 \tag{6}$$

$$1 + \tan^2 \theta = \sec^2 \theta \tag{7}$$

$$1 + \cot^2 \theta = \csc^2 \theta \tag{8}$$

These relations are invaluable for many considerations that follow in this book and should be memorized immediately. The first three relations have already been discussed (Section 8) and used. We shall prove only (1):

$$\frac{1}{\sin \theta} = \frac{1}{y/r} = \frac{r}{y} = \csc \theta$$

The proofs of (4) and (5) are similar. To prove (4):

$$\frac{\sin \theta}{\cos \theta} = \frac{y/r}{x/r} = \frac{y}{x} = \tan \theta$$

In order to prove (6), (7), and (8), we recall that for any angle θ in standard position, the coordinates of point P on its terminal side (see Figure 33) are related by the equation

$$x^2 + y^2 = r^2 \tag{9}$$

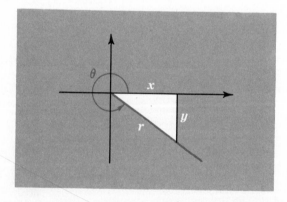

FIGURE 33

Dividing both sides by r^2, we get

$$\frac{x^2}{r^2} + \frac{y^2}{r^2} = \frac{r^2}{r^2}$$

$$\left(\frac{x}{r}\right)^2 + \left(\frac{y}{r}\right)^2 = 1$$

$$\cos^2\theta + \sin^2\theta = 1$$

Dividing Equation (9) in turn by x^2 and y^2, we get (7) and (8), respectively.

The student should be able to recognize these eight fundamental relations in other forms. For example,

(1) may be reduced to $\sin\theta \csc\theta = 1$.

(2) may be reduced to $\cos\theta = \dfrac{1}{\sec\theta}$.

(6) may be reduced to $\sin\theta = \pm\sqrt{1 - \cos^2\theta}$.

(6) may be reduced to $\cos\theta = \pm\sqrt{1 - \sin^2\theta}$.

(7) may be reduced to $\sec\theta = \pm\sqrt{1 + \tan^2\theta}$.*

The sign that should be chosen in the last three equations is determined by the quadrant in which θ lies. In Q I and Q IV, $\cos\theta = \sqrt{1 - \sin^2\theta}$; but in Q II and Q III, $\cos\theta = -\sqrt{1 - \sin^2\theta}$.

Again the student is warned against the careless habit of writing *sin* instead of *sin θ*. Equation (6) says that the square of the sine of any angle plus the square of the cosine of *that same angle* is equal to 1. It could just as well have been written

$$\sin^2 A + \cos^2 A = 1$$

or
$$\sin^2 7B + \cos^2 7B = 1$$

EXAMPLE 1 Prove or disprove:

$$\sin^4 5A + 2\sin^2 5A \cos^2 5A + \cos^4 5A = 1$$

---

*The student should realize from his study of algebra that $\sqrt{1 + \tan^2\theta}$ is *not* equal to $1 + \tan\theta$ for all θ.

SOLUTION   Since the given equation involves only sines and cosines, we shall attempt to derive it from (6), which says that for all values of θ,

$$\sin^2 \theta + \cos^2 \theta = 1$$

Square both sides of the equation:

$$(\sin^2 \theta + \cos^2 \theta)^2 = 1^2$$
$$\sin^4 \theta + 2 \sin^2 \theta \cos^2 \theta + \cos^4 \theta = 1$$

Now let $\theta = 5A$:

$$\sin^4 5A + 2 \sin^2 5A \cos^2 5A + \cos^4 5A = 1$$

This proves the statement is true for all values of A.

EXAMPLE 2   Prove or disprove:

$$\sin A + \cos A = 1$$

SOLUTION   We can demonstrate that this equation is not generally true by setting A equal to some specific angle and then showing that the two sides of the equation are not numerically equal. Choosing $A = 30°$, we have

$$\sin 30° + \cos 30° = 1$$
$$0.5 + 0.866 = 1$$
$$1.366 = 1 \qquad \text{False}$$

This proves conclusively that the given equation is not true for all values of A. It is, however, true for some values of A; e.g., if $A = 90°$, we have

$$\sin 90° + \cos 90° = 1$$
$$1 + 0 = 1 \qquad \text{True}$$

EXAMPLE 3   Simplify $\sqrt{\csc^2 3B - 1}$.

SOLUTION   Since $1 + \cot^2 3B = \csc^2 3B$, it follows that $\cot 3B = \pm \sqrt{\csc^2 3B - 1}$. This implies that $\sqrt{\csc^2 3B - 1} =$

$\pm \cot 3B$. If $3B$ is an angle in Q I or Q III, then $\cot 3B > 0$. In this case $\sqrt{\csc^2 3B - 1} = \cot 3B$. But if $3B$ is in Q II or Q IV, then $\cot 3B < 0$. By definition, $\sqrt{\csc^2 3B - 1}$ means the *principal* square root of $\csc^2 3B - 1$. It is positive or zero but never negative. In this case $\sqrt{\csc^2 3B - 1} = -\cot 3B$. Hence

$$\sqrt{\csc^2 3B - 1} = \cot 3B$$
$$\text{if } 3B \text{ is in Q I or Q III}$$

and $$\sqrt{\csc^2 3B - 1} = -\cot 3B$$
$$\text{if } 3B \text{ is in Q II or Q IV}$$

The student should note carefully that a general statement can be *disproved* by citing one instance in which it is not true. Such an instance is called a *counterexample*. But a general statement cannot be proved by merely showing that it is true for one special case. It must be proved for all cases.

EXERCISE 13

*Use the eight fundamental relations * to write each of the following expressions as a single trigonometric function of some angle.*

1. $\dfrac{1}{\tan 33°}$

2. $\dfrac{1}{\cot 7A}$

3. $\dfrac{\cos (A + B)}{\sin (A + B)}$

4. $\dfrac{\sin 88°}{\cos 88°}$

5. $\sqrt{\sec^2 A - 1}$

6. $-\sqrt{1 - \cos^2 250°}$

7. $\sqrt{1 - \sin^2 340°}$

8. $\sqrt{1 + \tan^2 10°}$

9. $-\sqrt{\csc^2 160° - 1}$

10. $\cot 77° \sin 77°$

Decide which of the following statements are valid consequences of the eight fundamental relations. If the statement is true, cite proof; if false, correct it.

11. $\dfrac{\sin 4A}{\cos 4A} = \tan A$

12. $5 \sec^2 \theta = \dfrac{1}{5 \cos^2 \theta}$

13. $\dfrac{64 \sin^3 A}{\cos^3 A} = (4 \tan A)^3$

14. $\sin \theta = \dfrac{1}{\sec \theta}$

---

*Do not express in terms of x, y, r.

15. $\tan^3 \theta \cot^3 \theta = 1$

16. $\dfrac{1}{\tan \theta} = \cot$

17. $\cot 2A = \dfrac{\cos 8A^2}{\sin 4A}$

18. $\tan^4 A = \dfrac{(\sin A)^4}{\cos^4 A}$

19. $1 + \cot \theta = \csc \theta$

20. $(3 \sin \theta)^2 = \dfrac{9}{\csc^2 \theta}$

21. $\cos 3A \csc 3A = 1$

22. $\cot^2 5B - \csc^2 5B = -1$

23. $(\tan^2 \theta - \sec^2 \theta)^3 = 1$

24. $\cot^4 \theta + 2 \cot^2 \theta + 1 = \csc^4 \theta$

25. $\sin \theta = -\sqrt{1 - \cos^2 \theta}$ holds for θ in Q II and Q III.

26. $\tan \theta = \sqrt{\sec^2 \theta + 1}$ holds for θ in Q I and Q III.

27. $\csc \theta = \sqrt{1 + \cot^2 \theta}$ holds for θ in Q I and Q II.

28. $\sin \theta = \sqrt{1 - \cos^2 \theta}$ holds for θ in Q I and Q IV.

29. $\sqrt{\sin^2 110°} = \sin 110°$

30. $\sqrt{\sin^2 220°} = \sin 220°$

31. $\csc 45° = \dfrac{1}{\sin 45°} = \dfrac{1}{\frac{1}{2} \sqrt{2}} = \dfrac{2}{\sqrt{2}} = \sqrt{2}$

32. $\cot 30° = \dfrac{\cos 30°}{\sin 30°} = \dfrac{\frac{1}{2} \sqrt{3}}{\frac{1}{2}} = \sqrt{3}$

25 ALGEBRAIC OPERATIONS WITH THE TRIGONOMETRIC FUNCTIONS

The expression $\sin \theta$, meaning the sine of angle θ, is an abstract number. It is the ratio of two distances, such as

$$\frac{2 \text{ ft}}{3 \text{ ft}} = \frac{2}{3} \quad \text{or} \quad \frac{2 \text{ cm}}{3 \text{ cm}} = \frac{2}{3}$$

For this reason it can be treated in the same way that we deal with numbers and letters (representing numbers) in algebra. For example, $\sin^3 \theta + \cos^3 \theta$ may be expressed as the sum of two cubes.* Since

$$a^3 + b^3 = (a + b)(a^2 - ab + b^2)$$
$$\sin^3 \theta + \cos^3 \theta = (\sin \theta + \cos \theta)(\sin^2 \theta - \sin \theta \cos \theta + \cos^2 \theta)$$
$$= (\sin \theta + \cos \theta)(1 - \sin \theta \cos \theta)$$

---

*Recall that $\sin^3 \theta$ is a short way of writing $(\sin \theta)^3$.

Also $(\sec \theta + \tan \theta)^2$ may be expanded as the square of a binomial to equal "the square of the first plus twice the product plus the square of the last":

$$(\sec \theta + \tan \theta)^2 = \sec^2 \theta + 2 \sec \theta \tan \theta + \tan^2 \theta$$

A glance at the fundamental relations reveals that the functions occurring most often are $\sin \theta$ and $\cos \theta$. Equations (1), (2), (4), and (5) express each of the other functions directly in terms of $\sin \theta$ and $\cos \theta$. Also, $\sin \theta$ and $\cos \theta$ are the only trigonometric functions that are defined for all θ. For these reasons it is advantageous in many problems to reduce an expression to sines and cosines.

EXAMPLE 1 Express $\dfrac{3 \csc \theta}{5 \csc \theta - 6 \cot^2 \theta}$ in terms of $\sin \theta$ and $\cos \theta$.

SOLUTION

$$\frac{3 \csc \theta}{5 \csc \theta - 6 \cot^2 \theta}$$

$$= \frac{3 \left(\dfrac{1}{\sin \theta} \right)}{5 \left(\dfrac{1}{\sin \theta} \right) - 6 \left(\dfrac{\cos^2 \theta}{\sin^2 \theta} \right)}$$

$$= \frac{\dfrac{3}{\sin \theta}}{\dfrac{5 \sin \theta - 6 \cos^2 \theta}{\sin^2 \theta}} \qquad \text{Getting a common denominator for the bottom}$$

$$= \frac{3}{\sin \theta} \cdot \frac{\sin^2 \theta}{5 \sin \theta - 6 \cos^2 \theta} \qquad \text{Inverting the denominator and multiplying}$$

$$= \frac{3 \sin \theta}{5 \sin \theta - 6 \cos^2 \theta} \qquad \text{Simplifying the fraction}$$

To express this quantity in terms of just $\sin \theta$, replace $\cos^2 \theta$ with $(1 - \sin^2 \theta)$ to obtain $\dfrac{3 \sin \theta}{5 \sin \theta - 6 + 6 \sin^2 \theta}$.

EXAMPLE 2 Express each of the other trigonometric functions of θ in terms of $\sin \theta$.

SOLUTION 1

$$\cos\theta = \pm\sqrt{1 - \sin^2\theta} \qquad\qquad \text{using (6)}$$

$$\tan\theta = \frac{\sin\theta}{\cos\theta} = \frac{\sin\theta}{\pm\sqrt{1-\sin^2\theta}} \qquad \text{using (4), (6)}$$

$$\cot\theta = \frac{\cos\theta}{\sin\theta} = \frac{\pm\sqrt{1-\sin^2\theta}}{\sin\theta} \qquad \text{using (5), (6)}$$

$$\sec\theta = \frac{1}{\cos\theta} = \frac{1}{\pm\sqrt{1-\sin^2\theta}} \qquad \text{using (2), (6)}$$

$$\csc\theta = \frac{1}{\sin\theta} \qquad\qquad \text{using (1)}$$

SOLUTION 2   Place θ in standard position, as in Figure 34. In order to make $y/r = \sin\theta$, let $y = \sin\theta$ and $r = 1$. The pythagorean theorem gives $x = \pm\sqrt{1 - \sin^2\theta}$. Then

$$\cos\theta = \frac{x}{r} = \pm\sqrt{1-\sin^2\theta}$$

$$\tan\theta = \frac{y}{x} = \frac{\sin\theta}{\pm\sqrt{1-\sin^2\theta}} \qquad \text{etc.}$$

EXERCISE 14

Simplify each of the following:

1. $\dfrac{\cos^4\theta - \sin^4\theta}{\cos^2\theta - \sin^2\theta}$

2. $\dfrac{\csc^3\theta + \cot^3\theta}{\csc\theta + \cot\theta}$

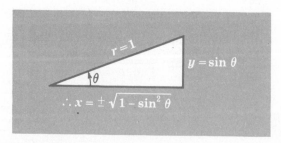

FIGURE 34

3. $\dfrac{\cot^2 \theta + 9 \cot \theta + 20}{\cot^2 \theta - 3 \cot \theta - 28}$ **4.** $\dfrac{\dfrac{1}{\cos \theta} - \dfrac{6}{\cos^2 \theta}}{\dfrac{1}{2 \cos^2 \theta} - \dfrac{3}{\cos^3 \theta}}$

Reduce each of the following to an expression that involves no function except $\sin \theta$ *and* $\cos \theta$. *Simplify.*

5. $\dfrac{\cot \theta}{\sqrt{\cot^2 \theta + 1}}$ **6.** $\dfrac{\sec \theta}{\tan \theta} + \dfrac{\tan \theta}{\cos \theta - \sec \theta}$

7. $\dfrac{\cot^2 \theta + \tan^2 \theta}{\csc \theta \cot^2 \theta + \sec \theta \tan \theta}$ **8.** $\dfrac{\cot \theta \cos \theta}{\csc \theta - \sin \theta}$

9. Express $\sec 224°$ in terms of $\tan 224°$.

10. Express $\tan 98°$ in terms of $\cot 98°$.

11. Express $\cos 112°$ in terms of $\sin 112°$.

12. Express $\cot 336°$ in terms of $\csc 336°$.

13. Express $\cos 5A$ in terms of $\sec 5A$.

14. Express $\sin 7B$ in terms of $\cos 7B$.

15. Express each of the other trigonometric functions in terms of $\tan \theta$. Why is your expression for $\sin \theta$ not valid when $\theta = 90°$?

16. Express each of the other trigonometric functions in terms of $\cos \theta$.

26 IDENTITIES AND CONDITIONAL EQUATIONS

An identity* is an equation that holds true for all permissible† values of the letters involved.

ILLUSTRATION 1 $x^2 - 9 = (x + 3)(x - 3)$ holds true for all values of x.

ILLUSTRATION 2 $x^2 + xy - 2y^2 = (x + 2y)(x - y)$ holds true for all values of x and y.

ILLUSTRATION 3 $x - \dfrac{x^2 - 7x}{x - 5} = \dfrac{2x}{x - 5}$ holds true for all permissible

---

*Also called an *identical equation.*

† The *permissible* values of the letters involved are all those values for which each side of the equation has meaning.

values of x, that is, for all values of x except $x = 5$. When $x = 5$, each side of the equation involves a fraction whose denominator is zero. Such fractions have no meaning, and we say their value does not exist.

ILLUSTRATION 4 $\sin^2 \theta + \cos^2 \theta = 1$ holds true for all values of θ.

ILLUSTRATION 5 $1 + \tan^2 \theta = \sec^2 \theta$ holds true for all permissible values of θ, that is, for all values of θ except 90°, 270°, and angles coterminal with them.

ILLUSTRATION 6 The following "trick with numbers" illustrates a simple identity.

> Choose any number except 0.
> Multiply your number by 5.
> To this number add the square of your original number.
> Multiply your result by 2.
> Divide the number you now have by the original number.
> Subtract 10.
> Divide by your original number.

If you have followed instructions, your result should be 2 regardless of your choice of the original number. To prove this, let x be the original number. Then the numbers that follow are $5x$, $5x + x^2$, $2(5x + x^2)$ or $x(10 + 2x)$, $10 + 2x$, $2x$, and 2. The identity used is

$$\frac{\dfrac{2(5x + x^2)}{x} - 10}{x} = 2$$

It holds for all values of x except 0. Try it for a fraction. For a negative number.

A conditional equation is an equation that *does not* hold true for all permissible values of the letters involved.

ILLUSTRATION 7 $2x - 7 = 3$ holds true for only one value of x, namely $x = 5$.

ILLUSTRATION 8 $x^2 - 8x + 15 = 0$ holds true for only two values of x, namely $x = 3$ and $x = 5$.

ILLUSTRATION 9 $x(x - 7)(x + 4) = 0$ holds true for only three values of x, namely $x = 0$, $x = 7$, and $x = -4$.

ILLUSTRATION 10 $\sin \theta = \cos \theta$ holds true for only two values of θ between $0°$ and $360°$. They are $\theta = 45°$ and $\theta = 225°$.

ILLUSTRATION 11 $\sin \theta = 5 + \cos \theta$ holds true for no value of θ.

The difference between an identity and a conditional equation can easily be seen from the contrasting definitions:

$$\left\{ \begin{matrix} An\ identity \\ A\ conditional\ equation \end{matrix} \right\} \ is\ an\ equation\ that \left\{ \begin{matrix} holds\ true \\ does\ not\ hold\ true \end{matrix} \right\}$$

for all permissible values of the letters involved.

An identity *says* that both sides of an equation are equal for all permissible values. The process by which we demonstrate that the two sides are identical is called "proving the identity." A conditional equation *asks*, "For what values of the unknowns is the left side of this equation equal to the right side?" The process by which these values are found is called "solving the equation."

27 TRIGONOMETRIC IDENTITIES

The eight fundamental relations are identities. By using them, we can prove other identities. For one who plans to go further in mathematics, or in subjects involving mathematics, it is highly important to gain a certain amount of experience in proving identities. For this reason we place considerable emphasis on the following examples and problems.

It is most desirable to prove an identity by *transforming one side* of the equation *to the other side, which should be left unaltered.* * The side with which

---

*The instructor may wish to permit the student to reduce each side of the equation independently to a common third expression. This method is sometimes desirable when both sides are quite complicated.

we work is usually the more complicated one. There is no set rule for making these transformations. The following suggestions will, however, indicate the first step in most cases.

1. If one side involves only one function of the angle, express the other side in terms of this function.

2. If one side is factorable, factor it.

3. If one side has only one term in its denominator (and several terms in its numerator), break up the fraction.

4. If one side contains one or more indicated operations (such as squaring an expression, adding fractions, or multiplying two expressions), begin by performing these operations. This is especially helpful if this side involves only sines and cosines.

5. When working with one side, keep an eye on the other side to see which transformation will most easily reduce the first side to the other side. It is frequently helpful to multiply the numerator and denominator of a fraction by the same expression. If possible, avoid introducing radicals.

6. When in doubt, express the more complicated side in terms of sines and cosines and then simplify.

At each step, look for some combination that can be replaced by a simpler expression.

The following examples illustrate the suggestions.

EXAMPLE 1 Prove the identity

$$3 \cos^4 \theta + 6 \sin^2 \theta = 3 + 3 \sin^4 \theta$$

PROOF

$$\begin{aligned} & 3 \cos^4 \theta + 6 \sin^2 \theta \\ & = 3(1 - \sin^2 \theta)^2 + 6 \sin^2 \theta \\ & = 3 - 6 \sin^2 \theta + 3 \sin^4 \theta + 6 \sin^2 \theta \\ & = 3 + 3 \sin^4 \theta \end{aligned} \qquad \bigg| \qquad 3 + 3 \sin^4 \theta$$

EXAMPLE 2 Prove the identity

$$\sec^2 \theta + \tan^2 \theta = \sec^4 \theta - \tan^4 \theta$$

PROOF $\sec^2 \theta + \tan^2 \theta$

$$\begin{aligned} & \sec^4 \theta - \tan^4 \theta \\ & = (\sec^2 \theta + \tan^2 \theta)(\sec^2 \theta - \tan^2 \theta) \\ & = (\sec^2 \theta + \tan^2 \theta) \cdot 1 \\ & = (\sec^2 \theta + \tan^2 \theta) \end{aligned}$$

EXAMPLE 3 Prove the identity

$$\frac{\sin \theta + \cot \theta}{\cos \theta} = \tan \theta + \csc \theta$$

PROOF

$$\frac{\sin \theta + \cot \theta}{\cos \theta} \qquad\bigg|\qquad \tan \theta + \csc \theta$$

$$= \frac{\sin \theta}{\cos \theta} + \frac{\cot \theta}{\cos \theta}$$

$$= \tan \theta + \frac{\dfrac{\cos \theta}{\sin \theta}}{\cos \theta}$$

$$= \tan \theta + \frac{1}{\sin \theta}$$

$$= \tan \theta + \csc \theta$$

EXAMPLE 4 Prove the identity

$$\frac{\sin \theta}{1 + \cos \theta} + \frac{1 + \cos \theta}{\sin \theta} = 2 \csc \theta$$

PROOF   The left side indicates the addition of two fractions that involve only sines and cosines of θ. We begin by adding these fractions.

$$\frac{\sin \theta}{1 + \cos \theta} + \frac{1 + \cos \theta}{\sin \theta} \qquad\bigg|\qquad 2 \csc \theta$$

$$= \frac{\sin^2 \theta + 1 + 2 \cos \theta + \cos^2 \theta}{(1 + \cos \theta) \sin \theta}$$

$$= \frac{2 + 2 \cos \theta}{(1 + \cos \theta) \sin \theta}$$

$$= \frac{2(1 + \cos \theta)}{(1 + \cos \theta) \sin \theta}$$

$$= \frac{2}{\sin \theta}$$

$$= 2 \csc \theta$$

EXAMPLE 5 Prove the identity

$$\frac{\cot \theta}{\csc \theta - 1} = \frac{\csc \theta + 1}{\cot \theta}$$

PROOF Since the numerator of the right side is $\csc \theta + 1$, let us multiply top and bottom of the left side by $\csc \theta + 1$ to get

$$\frac{\cot \theta \, (\csc \theta + 1)}{(\csc \theta - 1)(\csc \theta + 1)}$$

$$= \frac{\cot \theta \, (\csc \theta + 1)}{\csc^2 \theta - 1}$$

$$= \frac{\cot \theta \, (\csc \theta + 1)}{\cot^2 \theta}$$

$$= \frac{\csc \theta + 1}{\cot \theta} \qquad \text{which is the right side}$$

EXAMPLE 6 Prove the identity

$$\frac{\cot A + \csc B}{\tan B + \tan A \sec B} = \cot A \cot B$$

PROOF

$$\frac{\cot A + \csc B}{\tan B + \tan A \sec B} \qquad \Big| \qquad \cot A \cot B$$

$$= \frac{\dfrac{\cos A}{\sin A} + \dfrac{1}{\sin B}}{\dfrac{\sin B}{\cos B} + \dfrac{\sin A}{\cos A \cos B}}$$

$$= \frac{\dfrac{\sin B \cos A + \sin A}{\sin A \sin B}}{\dfrac{\sin B \cos A + \sin A}{\cos A \cos B}}$$

$$= \frac{\cos A \cos B}{\sin A \sin B}$$

$$= \cot A \cot B$$

The form illustrated in the preceding examples, in which a vertical line separates the two sides of the identity, is a convenient form to follow. It emphasizes the fact that, in trying to prove that the equation holds for all permissible values of the letters involved, we must *not* work with it as if it were an ordinary conditional equation. * The key point, however, is that each step in the transformation from the expression on one side of the identity to the expression on the other side must be clearly justified as an *identity transformation.*

EXERCISE 15

Prove each of the following identities by reducing one side to the other:

1. $\dfrac{1 - \dfrac{x - 7}{x + 7}}{\dfrac{1}{x + 7} - \dfrac{1}{x - 7}} = 7 - x$

2. $\dfrac{3 - \dfrac{2x - 11}{x - 2}}{2 - \dfrac{x^2 + 3}{x^2 + x - 6}} = \dfrac{x + 3}{x - 3}$

3. $\dfrac{7 - \dfrac{6y - 3}{y - 1}}{\dfrac{y}{y + 1} - \dfrac{y + 8}{y^2 - 1}} = \dfrac{y + 1}{y + 2}$

4. $\dfrac{1 - \dfrac{x + 22}{(x + 2)^2}}{x + 8 - \dfrac{x - 2}{x + 2}} = \dfrac{x - 3}{(x + 2)(x + 3)}$

---

*For example, to prove that $\cos \theta = \sin \theta / \tan \theta$, it would not be proper to multiply both sides by $\tan \theta$ to get $\cos \theta \tan \theta = \sin \theta$ and then to replace $\tan \theta$ with $\sin \theta / \cos \theta$ to get $(\cos \theta) \sin \theta / \cos \theta = \sin \theta$ or $\sin \theta = \sin \theta$. This does *not* prove that the given equation is an identity. It merely demonstrates that IF $\cos \theta = \sin \theta / \tan \theta$, THEN (it is necessary that) $\sin \theta = \sin \theta$. We could use this same improper procedure to prove that $3 = 5$ because multiplying both sides by 0 gives us $3(0) = 5(0)$, which is a true statement. Another incorrect procedure is to square both sides of an equation and then claim that the given equation is an identity because the squares of its two sides are identically equal. With this sort of reasoning, we could claim that $-4 = 4$ because $(-4)^2 = 4^2$. If $a^2 = b^2$, it does not necessarily follow that $a = b$.

5. $\dfrac{\tan \theta}{\sec \theta} = \sin \theta$

6. $\sin^2 A + \csc^2 B + \cos^2 A - \cot^2 B = 2$

7. $(1 - \sin^2 C)(1 + \tan^2 C) = 1$

8. $(1 + \cot^2 A) \cos^2 A = \cot^2 A$

9. $\dfrac{\csc B}{\sin B} - \dfrac{\cot B}{\tan B} = 1$

10. $\dfrac{\csc \theta - 1}{1 - \sin \theta} = \csc \theta$

11. $\sin \theta \cot \theta \sec \theta = 1$

12. $\cot B + \tan B = \csc B \sec B$

13. $7 \tan^2 \theta - 6 \sec^2 \theta + 8 = \sec^2 \theta + 1$

14. $\sec^4 \theta - 1 = \tan^4 \theta + 2 \tan^2 \theta$

15. $\cot^4 A + 3 \csc^4 A + 2 \csc^2 A = 4 \csc^4 A + 1$

16. $2 \cos^2 \theta + \sin^4 \theta = \cos^4 \theta + 1$

17. $\dfrac{\cot A + \csc A - \cos A}{\cot A \csc A} = \sin A + \tan A - \sin^2 A$

18. $\dfrac{1 - \cot C + 5 \sin C}{\tan C} = \cot C - \cot^2 C + 5 \cos C$

19. $\dfrac{3 \sin \theta + 8 \csc \theta}{4 \sin \theta} = \dfrac{3}{4} + 2 \csc^2 \theta$

20. $\dfrac{\cos \theta + 4 \sec \theta + \sin^2 \theta \sec \theta}{\sec \theta} = 5$

21. $\dfrac{7 \sec^4 \theta - 7 \tan^4 \theta}{\sec^2 \theta + \tan^2 \theta} = 7$

22. $\dfrac{\csc^3 \theta + \sin^3 \theta}{\csc \theta + \sin \theta} = \cot^2 \theta + \sin^2 \theta$

23. $\dfrac{\cot^3 B - \tan^3 B}{\cot B - \tan B} = \sec^2 B + \csc^2 B - 1$

24. $(\sin \theta - \cos \theta)(\sec \theta + \csc \theta) = \tan \theta - \cot \theta$

25. $\dfrac{\tan A + \tan B}{1 - \tan A \tan B} = \dfrac{\cot A + \cot B}{\cot A \cot B - 1}$

26. $\dfrac{\cos A + \cos B}{\sec A + \sec B} = \cos A \cos B$

27. $\sec A \csc B (\cot A + \tan B) = \csc A \csc B + \sec A \sec B$

28. $\dfrac{\sec^2 C - 2}{\tan^2 C + 9 \tan C + 8} = \dfrac{\tan C - 1}{\tan C + 8}$

29. $\dfrac{\sin^2 \theta + 8}{\sin^2 \theta + \cos \theta + 11} = \dfrac{3 - \cos \theta}{4 - \cos \theta}$

30. $\dfrac{1}{\csc \theta - \sin \theta} = \tan \theta \sec \theta$

31. $\theta \cos^2 \theta + \sec \theta + \theta \sin^2 \theta - \tan \theta \sin \theta = \theta + \cos \theta$

32. $\sin \theta + \cos \theta \cot \theta = \csc \theta$

In Problems 33 to 40, either prove or disprove the statement. (Prove that the equation is, or is not, an identity.)

33. $\dfrac{1 + \cos C}{1 - \cos C} - \dfrac{1 - \cos C}{1 + \cos C} = 4 \cos C$

34. $(\csc \theta + 3 \sin \theta)^2 - (\csc \theta - 3 \sin \theta)^2 = 6$

35. $\dfrac{\cos \theta}{1 + \sin \theta} + \dfrac{1 + \sin \theta}{\cos \theta} = 2 \sec \theta$

36. $\dfrac{1 + \sec B}{\tan B} + \dfrac{\tan B}{1 - \sec B} = 0$

37. $\dfrac{\tan \theta - \cos \theta}{\cos \theta} = \dfrac{\sin^2 \theta + \sin \theta - 1}{\cos^2 \theta}$

38. $\csc A + \sin A = \dfrac{2 - \cos^2 A}{\sin A}$

39. $\dfrac{1}{1 + \sin \theta} + \dfrac{1}{1 - \sin \theta} = 2$

40. $(3 \sin \theta + 4 \cos \theta)^2 + (4 \sin \theta - 3 \cos \theta)^2 = 25$

In Problems 41 to 44, assume the angle is in Q I. Prove each identity.

41. $\sqrt{\dfrac{1 + \cos \theta}{1 - \cos \theta}} = \dfrac{\sin \theta}{1 - \cos \theta}$ *Hint:* Rationalize the left side by

multiplying under the radical by $\dfrac{1 - \cos \theta}{1 - \cos \theta}$.

42. $\sqrt{\dfrac{\sec \theta - 1}{\sec \theta + 1}} = \csc \theta - \cot \theta$

43. $\sqrt{\dfrac{\csc \theta - \cot \theta}{\csc \theta + \cot \theta}} = \dfrac{1 - \cos \theta}{\sin \theta}$

44. $\sqrt{\dfrac{\sec \theta + \tan \theta}{\sec \theta - \tan \theta}} = \dfrac{1}{\sec \theta - \tan \theta}$

45. State four nonpermissible values of A for which the identity in Problem 17 does not hold true.

46. State three nonpermissible values of θ for which the identity in Problem 10 does not hold true.

47. State six nonpermissible values of B for which the identity in Problem 23 does not hold true.

48. State two nonpermissible values of A for which the identity in Problem 8 does not hold true.

49. Verify the identity in Problem 25 for $\tan A = 3$, $\tan B = \frac{1}{5}$.

50. Verify the identity in Problem 30 for $\csc \theta = 4$ with θ in Q II.

51. Verify the identity in Problem 43 for $\sin \theta = \frac{5}{13}$ with θ in Q I.

52. Verify the identity in Problem 12 for $B = 30°$.

Prove each of the following identities.

53. $\dfrac{5 \sec A - \tan A}{8 \sec A + \tan A} + \dfrac{5 \csc A + 1}{8 \csc A - 1} = \dfrac{82 - 2 \cos^2 A}{63 + \cos^2 A}$

54. $\tan \theta = (\sec \theta + 1)(\csc \theta - \cot \theta)$

55. $\dfrac{\tan C + \tan A}{\cot C + \cot A} + \dfrac{1 - \tan C \tan A}{1 - \cot C \cot A} = 0$

56. $(\sin A \sin B - \cos A \cos B)^2 + (\sin A \cos B + \cos A \sin B)^2 = 1$

57. $\sec \theta (\sin \theta - 1)(\tan \theta + \sec \theta) = -1$

58. $\dfrac{1 - \cos \theta}{1 + \cos \theta} = \cot^2 \theta - 2 \csc \theta \cot \theta + \csc^2 \theta$

59. $\dfrac{\tan^2 \theta}{1 + \csc^2 \theta + \tan^2 \theta} = \sin^4 \theta$

60. $\dfrac{\cot^2 A}{\sin^2 B} - \dfrac{\cot^2 B}{\sin^2 A} = \cot^2 A - \cot^2 B$

*61. $\csc \theta + \cot \theta + 1 = \dfrac{2 \csc \theta \cot \theta + 2 \cot^2 \theta}{\csc \theta + \cot \theta - 1}$

*62. $\dfrac{\sin B + \cos B - 1}{\sin B - \cos B + 1} = \dfrac{1 - \sin B}{\cos B}$

---

*See the suggestion in Example 5, Section 27.

63. $(\cot\theta - \tan\theta)^2 = \cot^2\theta(2 - \sec^2\theta)^2$

64. $\csc^6\theta - \cot^6\theta = 1 + 3\csc^2\theta\cot^2\theta$

65. What values of a, b, and c will make the following equation an identity?

$$3 + 4\cos^2\theta + 5\cos^4\theta = a + b\sin^2\theta + c\sin^4\theta$$

66. Show that $\frac{1}{20}\sec^4 5\theta + \frac{1}{40}\sec^2 5\theta$ is identically equal to $\frac{1}{20}\tan^4 5\theta + \frac{1}{8}\tan^2 5\theta + C$, where C is a constant. Determine the value of C.

4

RELATED ANGLES

28 RELATED ANGLES

The fact that the table of trigonometric functions deals with only acute angles should have implied that functions of larger angles are expressible in terms of functions of acute angles. Such is really the case. Since coterminal angles have the same trigonometric functions, we shall consider only those angles between 0° and 360°. In order to determine the functions of angles larger than 90°, we introduce the concept of the related angle.

> The related angle of a given angle θ is the positive acute angle between the x axis and the terminal side of θ.

Hence (see Figure 35), *to find the related angle of θ*

If θ is in Q II, *subtract θ from* 180°.
If θ is in Q III, *subtract* 180° *from θ.*
If θ is in Q IV, *subtract θ from* 360°.

It is to be noted that, in finding the related angle, we always work to or from 180° or 360°, never 90° or 270°. Thus

The related angle of 160° is 20°.
The related angle of 260° is 80°.
The related angle of 310° is 50°.
The related angle of 500° is the related angle of (500° − 360°) = 140°, which is 40°.

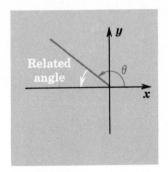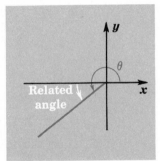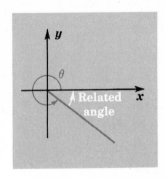

FIGURE 35

29 REDUCTION TO FUNCTIONS OF AN ACUTE ANGLE

Two numbers are said to be numerically equal if they are equal, except perhaps for sign. If a and b are numerically equal,* then $a = \pm b$.

RELATED-ANGLE THEOREM   Any trigonometric function of an angle is *numerically* equal to the same function of its related angle.

To prove this, let θ_1 be any positive acute angle, and let θ_2, θ_3, θ_4 be positive angles, one in each of the other three quadrants, such that their common related angle is θ_1 (see Figure 36). Choose points P_1, P_2, P_3, P_4 on the terminal sides of these angles so that all four points have the same radius vector r. The four triangles of reference are congruent. Therefore, the corresponding sides are numerically equal. Hence

$$x_2 = -x_1 \qquad x_3 = -x_1 \qquad x_4 = x_1$$

and

$$y_2 = y_1 \qquad y_3 = -y_1 \qquad y_4 = -y_1$$

Then

$$\sin \theta_2 = \frac{y_2}{r} = \frac{y_1}{r} = \sin \theta_1$$

$$\sin \theta_3 = \frac{y_3}{r} = \frac{-y_1}{r} = -\frac{y_1}{r} = -\sin \theta_1$$

$$\sin \theta_4 = \frac{y_4}{r} = \frac{-y_1}{r} = -\frac{y_1}{r} = -\sin \theta_1$$

---

*They have the same *absolute value*.

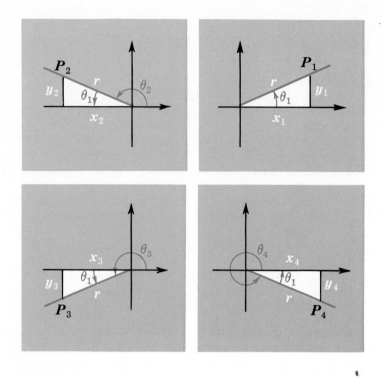

FIGURE 36

and
$$\cos \theta_2 = \frac{x_2}{r} = \frac{-x_1}{r} = -\frac{x_1}{r} = -\cos \theta_1$$

$$\cos \theta_3 = \frac{x_3}{r} = \frac{-x_1}{r} = -\frac{x_1}{r} = -\cos \theta_1$$

$$\cos \theta_4 = \frac{x_4}{r} = \frac{x_1}{r} = \cos \theta_1$$

Similarly, each of the other functions of θ_2, θ_3, and θ_4 is *numerically* equal to the same function of θ_1, the common related angle. The proper sign, $+$ or $-$, is determined by the quadrant in which the given angle lies.

EXAMPLE 1 Use the related-angle theorem to find sin 120° and cos 120° without tables.

SOLUTION The related angle of 120° is 180° − 120° = 60°. Since 120° is in Q II, its sine is positive and its cosine is negative. Hence

$$\sin 120° = +\sin 60° = \frac{\sqrt{3}}{2}$$

$$\cos 120° = -\cos 60° = -\tfrac{1}{2}$$

The remaining functions can be found by using the fundamental identities:

$$\tan 120° = \frac{\sin 120°}{\cos 120°} = \frac{\sqrt{3}/2}{-\frac{1}{2}} = -\sqrt{3}$$

$$\cot 120° = \frac{1}{\tan 120°} = -\frac{1}{\sqrt{3}} = \frac{-\sqrt{3}}{3} \qquad \text{etc.}$$

EXAMPLE 2 Use the related-angle theorem to find $\sin 255°$ and $\tan 255°$.

SOLUTION   The related angle of $255°$ is $255° - 180° = 75°$. Hence

$$\sin 255° = -\sin 75° = -0.9659$$
$$\tan 255° = +\tan 75° = 3.732$$

EXERCISE 16

Use the related-angle theorem to find the exact values of the sine and cosine of the following angles without using a calculator or tables.

1. 210°	**2.** 135°	**3.** 315°	**4.** 330°
5. 300°	**6.** 240°	**7.** 150°	**8.** 225°
9. 855°	**10.** 1050°	**11.** 600°	**12.** 480°

Prove or disprove the following statements without using a calculator or tables.

13. $\cos^2 136° + \sin^2 44° = 1$ **14.** $\csc^2 333° = 1 + \cot^2 27°$

15. $\sec 95° = \sec 85°$ **16.** $\cos 162° = -\sin 72°$

17. $\sin 255° > \sin 250°$ **18.** $\sin^2 200° + \sin^2 70° = 1$

19. $\dfrac{\cos 111°}{\cos 339°} = -\cot 69°$ **20.** $\dfrac{\sin 140°}{\cos 220°} = \tan 40°$

Name one angle in each of the other three quadrants whose trigonometric functions are numerically equal to those of the given angle.

21. 262°   **22.** 339°   **23.** 48°   **24.** 130°

Use the related-angle theorem and tables to find the values of the following functions.
(If you use a calculator, your instructor may wish to give you additional practice in using it by assigning some of these problems.)

25. tan 239°   **26.** cot 320°   **27.** tan 173°

28. sin 218°   **29.** cos 337.8°   **30.** sin 104.7°

31. cos 256.2°   **32.** tan 293.4°   **33.** cot 129.03°

34. cos 184.58°   **35.** sin 341.25°   **36.** cos 119.46°

30 TRIGONOMETRIC FUNCTIONS OF $(-\theta)$

Let θ be any angle. Then $(-\theta)$ indicates the same amount of rotation but in the *opposite* direction. Place both angles in standard position on the same coordinate system. (See Figure 37.) Choose any point $P(x,y)$ on the terminal side of θ. Drop a perpendicular from P to the x axis and extend it until it strikes the terminal side of $(-\theta)$ at $P'(x,y')$. The triangles of reference, OPM and $OP'M$, are congruent. Why? Hence $OP = OP' = r$. But y and y' are only numerically equal. Since $y' = -y$, we conclude that

$$\sin(-\theta) = \frac{y'}{r} = \frac{-y}{r} = -\frac{y}{r} = -\sin\theta$$

$$\cos(-\theta) = \frac{x}{r} = \cos\theta$$

$$\tan(-\theta) = \frac{y'}{x} = \frac{-y}{x} = -\frac{y}{x} = -\tan\theta$$

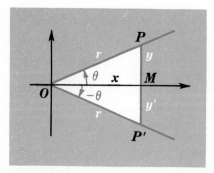 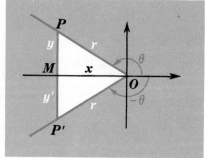

FIGURE 37

Similarly,

$$\cot(-\theta) = -\cot\theta$$
$$\sec(-\theta) = \sec\theta$$
$$\csc(-\theta) = -\csc\theta$$

The student should draw figures for θ in the other quadrants and also for θ a negative angle. For all possible positions of θ the following is true.

THEOREM   For any angle θ,

$$\sin(-\theta) = -\sin\theta$$
$$\cos(-\theta) = \cos\theta$$
$$\tan(-\theta) = -\tan\theta$$

The other three functions behave as do their reciprocals.

EXAMPLE   Compute $\sin(-225°)$, $\cos(-225°)$, and $\tan(-225°)$.

SOLUTION   Using the preceding theorem and the related-angle theorem, we obtain

$$\sin(-225°) = -\sin 225° = -(-\sin 45°) = \frac{\sqrt{2}}{2}$$

$$\cos(-225°) = \cos 225° = -\cos 45° = -\frac{\sqrt{2}}{2}$$

$$\tan(-225°) = -\tan 225° = -\tan 45° = -1$$

A function $f(\theta)$ is said to be an *even* function if, for all permissible values of θ, $f(-\theta) = f(\theta)$.

Examples of even functions* of θ are: $7\theta^2$, $\cos\theta$, $5\theta^6 - \theta^2$, and 1.

A function $f(\theta)$ is called an *odd* function if, for all permissible values of θ, $f(-\theta) = -f(\theta)$.

---

*See the second footnote in Section 83.

Examples of odd functions of θ are $8\theta^3$, $\sin\theta$, $\tan\theta$, and $-\theta^5$. Some functions, such as $\theta^5 + \theta^4$, are neither even nor odd.

EXERCISE 17

Use the theorem in Section 30 to compute the exact values of the sine, cosine, and tangent of each of the following angles. Do not use a calculator or tables.

1. $-45°$ **2.** $-60°$ **3.** $-90°$ **4.** $-30°$

Prove or disprove the following statements without using a calculator or tables.

5. $\sin(-110°) = \sin 70°$ **6.** $\cos(-181°) = \cos 1°$

7. $\dfrac{\cos(-130°)}{\cos(-220°)} = \tan 40°$ **8.** $\dfrac{\sin(-\theta)}{\cos(-\theta)} = -\tan(-\theta)$

9. $\tan^2(-\theta) = \sec^2\theta - 1$ **10.** $\cos^2\theta - \sin^2(-\theta) = 1$

11. $\cot(-\theta)\tan(-\theta) = -1$ **12.** $\csc 57° \sin(-57°) = -1$

13. $\cos(-250°)$ is a positive number.

14. $\tan(-161°)$ is a positive number.

15. $\sin(-202°)$ is a negative number.

16. Draw a figure and prove that, if θ is a negative angle in Q II, then
(*a*) $\sin(-\theta) = -\sin\theta$, (*b*) $\cos(-\theta) = \cos\theta$.

17. Draw a figure and prove that, if θ is a positive angle in Q IV, then
(*a*) $\sin(-\theta) = -\sin\theta$, (*b*) $\cos(-\theta) = \cos\theta$.

18. Identify each of the following functions as an even function, an odd function, or neither.
(*a*) $\cos 3\theta$ (*b*) $\tan(-6\theta)$ (*c*) $\theta^5 - 7$ (*d*) $3\theta^7 - 4\theta^3$
(*e*) $\tan^2\theta$ (*f*) $\sin^5\theta$ (*g*) $5\theta^8 + 6\theta^2$ (*h*) $\cos\theta - \tan\theta$

5 RADIAN MEASURE

31 THE RADIAN

Thus far we have employed the *degree* as the unit of measure for angles. It may be thought of as $\frac{1}{360}$ of the angular magnitude about a point. For many practical purposes, the degree is a convenient unit, but most of the applications of the trigonometric functions which require higher mathematics, particularly calculus, are simplified if another unit, the radian, is used. For this reason, students planning to study calculus should make a special effort to learn radian measure thoroughly and to use it, so that it will be familiar to them when they need it.

> A radian is an angle which, if its vertex is placed at the center of a circle, intercepts an arc equal in length to the radius of the circle.

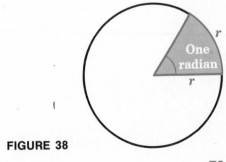

FIGURE 38

32 RADIANS AND DEGREES

According to the definition of a radian, the number of radians in a circle is equal to the number of times the radius can be laid off along the circumference (Figure 39). Since $c = 2\pi r$, the number of radians in a circle is $2\pi r/r = 2\pi$. But the number of degrees in a circle is $360°$. Hence

$$2\pi \text{ radians} = 360°$$

or $$\pi \text{ radians} = 180° \tag{1}$$

Hence $$1 \text{ radian} = \frac{180°}{\pi} \doteq \frac{180°}{3.14159} \tag{2}$$

$$\doteq 57.2958° \doteq 57° \ 17.75'$$

where $60' = 60$ minutes $= 1°$.

The symbol $\doteq$ means "is approximately equal to." (It is worth observing that 1 radian is a little less than $60°$.)

Also, $$1° = \frac{\pi}{180} \text{ radians} \doteq 0.017453 \text{ radian} \tag{3}$$

It is better not to try to learn Equations (2) and (3). Memorize (1) and derive your results from it. In expressing the more common angles in terms of radians, we usually leave the result in terms of π. Thus $90° = \pi/2$. *When no unit of measure is indicated, it is understood that an angle is expressed in radian.* Thus if we read $\theta = \pi/6$, we understand that $\theta = \pi/6$ radian $= 30°$.

The student should become quite familiar with the radian measure of $30°$, $45°$, $60°$, and $90°$, and the angles that are related to them, namely $120°$, $135°$, $150°$, $180°$, $210°$, $225°$, $240°$, $270°$, $300°$, $315°$, and $330°$.

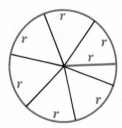

FIGURE 39

EXAMPLE 1 Express 12° 34′ 55″ in decimal degrees to the nearest thousandth of a degree.

SOLUTION Inasmuch as $60' = 1°$ and $60'' = 60 \text{ seconds} = 1'$,

$$34' \, 55'' = \left(\frac{34}{60} + \frac{55}{3600} \right)^° = \frac{2095°}{3600} \doteq 0.5819°$$

Rounded off, $12° \, 34' \, 55'' = 12.582°$

CALCULATOR SOLUTION Using a four-function $(+, -, \times, \div)$ calculator, we convert the 55 seconds to minutes; then add this to 34 minutes, convert the result to degrees, and finally add this to 12°. With an AOS, we press

$\boxed{55}\ \boxed{\div}\ \boxed{60}\ \boxed{=}^*\ \boxed{+}\ \boxed{34}\ \boxed{=}\ \boxed{\div}\ \boxed{60}\ \boxed{=}^*\ \boxed{+}\ \boxed{12}\ \boxed{=}$

With RPN, we press

$\boxed{55}\ \boxed{\text{ENTER}}\ \boxed{60}\ \boxed{\div}\ \boxed{34}\ \boxed{+}\ \boxed{60}\ \boxed{\div}\ \boxed{12}\ \boxed{+}$

In each case the displayed result is 12.58194444.

Henceforth, in this book we shall usually express angles in decimal degrees rather than in degrees, minutes, and seconds.

EXAMPLE 2 Express $\dfrac{7\pi}{9}$ in terms of degrees.

SOLUTION Since
$$\pi = 180°$$
$$\tfrac{7}{9}\pi = \tfrac{7}{9}(180°) = 140°$$

EXAMPLE 3 Express 6.75° in terms of radians.

---

*On many calculators these steps may be omitted.

SOLUTION 1 Since $180° = \pi$

$$1° = \frac{\pi}{180}$$

$$6.75° = 6.75\left(\frac{\pi}{180}\right) = \frac{675\pi}{18,000} = \frac{3\pi}{80}$$

This is the exact result. If π is replaced by 3.1416, we get

$$6.75° = 0.1178$$

correct to four decimal places.

SOLUTION 2 Use Table 1 and interpolate:

$$6.70° = 0.1169$$
$$6.75° =$$
$$6.80° = 0.1187$$

Hence $6.75° = 0.1178$

CALCULATOR SOLUTION   If, in addition to a degree-radian *switch*, your calculator has a degree/radian (D/R) *key*, set the *switch* for radians. Then press

$$\boxed{6.75}\ \boxed{\text{D/R}}$$

The displayed result is $0.1178097245 \rightarrow 0.1178.$ $\Bigg($If the degree-radian switch is set for $\begin{Bmatrix}\text{degrees}\\\text{radians}\end{Bmatrix}$, pressing the $\boxed{\text{D/R}}$ key instructs the calculator to convert the displayed angle from $\begin{Bmatrix}\text{radians}\\\text{degrees}\end{Bmatrix}$ to $\begin{Bmatrix}\text{degrees}\\\text{radians}\end{Bmatrix}.\Bigg)$

EXAMPLE 4 Express 2 radians in terms of degrees.

SOLUTION 1   Since $\pi \text{ radians} = 180°$

$$1 \text{ radian} = \frac{180°}{\pi}$$

$$2 \text{ radians} = \frac{360°}{\pi}$$

This exact result may be approximated by the decimal form $\dfrac{360°}{3.1416} \doteq 114.59°$, which is correct to a hundredth of a degree.

SOLUTION 2   Since

$$1 \text{ radian} = 57.2958°$$
$$2 \text{ radians} = 2(57.2958°)$$
$$= 114.5916° \to 114.59°$$

CALCULATOR SOLUTION   With the degree-radian *switch* set for degrees, press

$$\boxed{2}\ \boxed{\text{D/R}}$$

The displayed result is 114.591559. Hence, to two-decimal-place accuracy,

$$2 \text{ radians} = 114.59°$$

EXAMPLE 5   Evaluate $\csc\dfrac{7\pi}{6}$.

SOLUTION
$$\csc\frac{7\pi}{6} = \csc\frac{7(\overset{30°}{\cancel{180°}})}{\cancel{6}}$$

$$= \csc 210° = \frac{1}{\sin 210°} = \frac{1}{-\sin 30°}$$

$$= \frac{1}{-\frac{1}{2}} = -2$$

CALCULATOR SOLUTION   Set the switch for radians. With an AOS, press

$$\boxed{7}\ \boxed{\times}\ \boxed{\pi}\ \boxed{=}\ \boxed{\div}\ \boxed{6}\ \boxed{=}\ \boxed{\sin}\ \boxed{1/x}$$

to get the result, -2. With RPN, press

$$\boxed{7}\ \boxed{\text{ENTER}}\ \boxed{\pi}\ \boxed{\times}\ \boxed{6}\ \boxed{\div}\ \boxed{\sin}\ \boxed{1/x}$$

to get -2.

EXERCISE 18

Express the following in decimal degrees to the nearest thousandth of a degree.

1. $47°14'30''$ **2.** $83° 52' 18''$ **3.** $29° 50' 0''$ **4.** $76° 29' 45''$

Express in degrees. (The given angle is understood to be in radians.) Do not use a calculator or tables.

5. $\dfrac{3\pi}{10}$ **6.** $\dfrac{5\pi}{18}$ **7.** $-\dfrac{\pi}{20}$ **8.** $\dfrac{7\pi}{3}$

9. $\dfrac{17\pi}{18}$ **10.** $\dfrac{4\pi}{5}$ **11.** $\dfrac{14\pi}{9}$ **12.** $\dfrac{5\pi}{2}$

13. $\dfrac{2\pi}{9}$ **14.** 6π **15.** $\dfrac{\pi}{8}$ **16.** $\dfrac{11\pi}{10}$

Convert from radians to degrees; write the result to the nearest hundredth of a degree.

17. 6.3 **18.** -0.9 **19.** 1.01 **20.** 4.8

Express in radians, leaving the result in terms of π. Do not use a calculator.

21. $330°$ **22.** $225°$ **23.** $120°$ **24.** $-60°$
25. $45°$ **26.** $150°$ **27.** $210°$ **28.** $315°$
29. $1800°$ **30.** $400°$ **31.** $495°$ **32.** $900°$

Convert to radian measure, obtaining the result to four decimal places.

33. $50.50°$ **34.** $33.33°$ **35.** $9.08°$ **36.** $200.30°$

Find the exact values of the following without using tables. (The instructor may wish to permit the use of a calculator.)

37. $\sin\dfrac{5\pi}{6}$ **38.** $\cos\dfrac{\pi}{3}$ **39.** $\cos\dfrac{\pi}{4}$ **40.** $\sin\dfrac{5\pi}{4}$

41. $\cos\dfrac{4\pi}{3}$ **42.** $\sin\dfrac{7\pi}{4}$ **43.** $\sin\dfrac{\pi}{6}$ **44.** $\cos\left(-\dfrac{7\pi}{6}\right)$

45. $\tan\dfrac{3\pi}{4}$ **46.** $\tan\left(-\dfrac{11\pi}{6}\right)$ **47.** $\csc\dfrac{5\pi}{3}$ **48.** $\sec\dfrac{2\pi}{3}$

49. $\csc\left(-\dfrac{\pi}{2}\right)$ **50.** $\sec \pi$ **51.** $\cot\dfrac{3\pi}{2}$ **52.** $\tan 2\pi$

Use a calculator or Table 1 to evaluate the following to four decimal places. (The angle is understood to be in radians.)

53. $\cos 1.3456$ **54.** $\tan 0.4747$ **55.** $\sin 0.7383$ **56.** $\cos 0.0733$

In Problems 57 to 60, leave the result in terms of π.

57. Through how many radians does the minute hand of a clock rotate in 25 min? In 9 h?

58. One angle of a triangle is $5\pi/12$. Another angle is $63°$. Express the third angle in radians.

59. Express in radians the smaller angle made by the hands of a clock at 7:00. At 4:30. At 12:45.

60. Through how many radians does the hour hand of a clock rotate in 3 h? In 5 days?

33 LENGTH OF A CIRCULAR ARC

In a circle of radius r, let an arc of length s be intercepted by a central angle θ. Then

$$s = r\theta \qquad \text{where } \theta \text{ is in radians}$$

To prove this, recall that in any circle the arc intercepted by a central angle is proportional to this angle. Hence if 1 radian intercepts an arc equal to the radius, then θ radians intercept an arc equal to θ times the radius. It is to be remembered that *the central angle must be measured in radians.*

In the equation $s = r\theta$, if two of the quantities s, r, θ are known, the third can be found.

EXAMPLE   A circle has a radius of 100 cm. (*a*) How long is the arc intercepted by a central angle of $72°$? (*b*) How large is the central angle that intercepts an arc of 30 cm?

SOLUTION* (*a*) Since

$$180° = \pi \text{ radians}$$

$$1° = \frac{\pi}{180} \text{ radian}$$

$$72° = 72 \cdot \frac{\pi}{180} \text{ radians} = \frac{2\pi}{5} \text{ radians}$$

---

*In all discussions and problems in this chapter, all figures are to be considered as exact. They are not approximations. For the sake of uniformity, write all approximate results with three-figure accuracy. It may be convenient to use $1/\pi = 0.3183$.

Using

$$s = r\theta \qquad \text{where } \theta \text{ is in radians}$$

$$s = 100 \cdot \frac{2\pi}{5} = 40\pi \text{ cm (or } 126 \text{ cm)}$$

(*b*) Using

$$s = r\theta$$

$$30 = 100\theta$$

$$\theta = \frac{30}{100} = \frac{3}{10} \text{ radian} \quad \left(\text{or } \frac{3}{10} \cdot \frac{180°}{\pi} = \frac{54°}{\pi} \doteq 17.2° \right)$$

The color results are exact; the italicized answers are merely three-figure approximations.

34 THE WRAPPING FUNCTION

Up to this point, most of our discussions have dealt with the trigonometric functions of *angles*. For example, the sine function was defined as $\sin = \{(\theta, \sin \theta)\}$, where θ is a certain number of degrees. We shall now show that the six trigonometric functions may be defined as functions of real numbers—with no reference whatsoever to angles or triangles. To this end we introduce the *wrapping* function.

Consider the graph of a unit circle (radius 1) with center at the origin (Figure 40). The equation of this circle is $x^2 + y^2 = 1$. We intend to show that every real number can be associated with a unique point on the circumference of this circle. We begin by drawing a real-number line (the *s* axis) tangent to the unit circle at $A(1, 0)$, using OA as the unit and A as the zero point, and making the positive direction upward (Figure 40). Now imagine that this real-number line is completely flexible (like a tape measure). Think of wrapping the *positive* part of the *s* axis in a *counterclockwise* direction about the unit circle, and wrapping the negative part about the circle in a clockwise direction (Figure 41).

Obviously the number 0 on the *s* axis is associated with the point A on the unit circle. Using $c = 2\pi r$, we find the circumference of the circle to be 2π. Hence the real number $s = \dfrac{\pi}{2}$ (which is $\frac{1}{4}$ of the circle's circum-

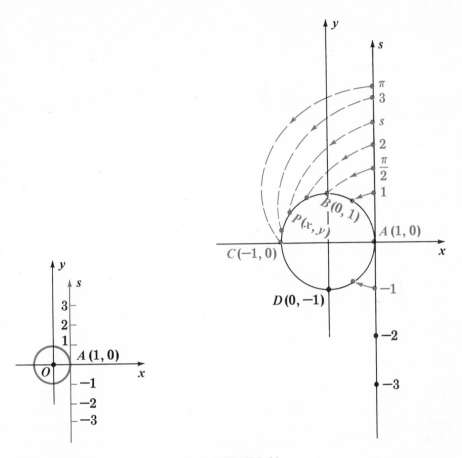

FIGURE 40   **FIGURE 41**

ference) is associated with the point $B(0, 1)$, the real number π is associated with the point $C(-1, 0)$, and the number $\dfrac{3\pi}{2}$ is associated with the point $D(0, -1)$. Because the circle's circumference is 2π, the portion of the s axis from 0 to 2π will wrap the circle one time, the portion of the s axis from 2π to 4π will wrap the circle a second time, etc. Each of the numbers 0, 2π, 4π, 6π, etc., is associated with the point $A(1, 0)$. In general, if the real number s is associated with the point P, then all the real numbers $s + 2\pi k$, where k is an integer (such as $s + 2\pi$, $s + 4\pi$, . . . , $s - 2\pi$, $s - 4\pi$, . . .), are associated with P. However, each positive real number is associated with one and only one point on the unit circle.

In similar fashion, wrapping the negative part of the s axis in a clockwise direction about the unit circle enables us to associate each negative number with a unique point on the circle.

We define the wrapping function as

$$W = \{[s, (x, y)]\}$$

where s is any real number and x and y are the coordinates of the associated point on the unit circle.

35 THE CIRCULAR FUNCTIONS

If s is a real number that is associated with the point $P(x, y)$ on the unit circle, the *circular* functions are defined as follows:

$$\sin s = y \qquad \cos s = x \qquad \tan s = \frac{y}{x}$$

$$\cot s = \frac{x}{y} \qquad \sec s = \frac{1}{x} \qquad \csc s = \frac{1}{y}$$

Thus the circular functions are defined as functions of real numbers.

Because the s axis is wrapped about the unit circle, we know that if the real number s is associated with the point $P(x, y)$ on the unit circle (Figure 42), then the length of the arc from A to P is s. In order to establish a relationship between the circular functions and the trigonometric functions of an angle expressed in radians, we shall use the equation $s = r\theta$, θ in radians, to observe that the radian measure of angle AOP is $s/r = s/1 = s$. Using the definitions of the trigonometric func-

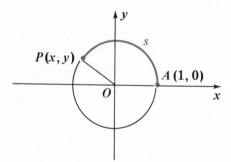

FIGURE 42

tions of a general angle (Section 7), we find that

$$\text{The sine of angle } AOP = \frac{y}{r} = \frac{y}{1} = y = \sin s$$

Hence $\sin s$ (the sine of the real number s) $= \sin s$ (radians). Similar statements can be made for the other five trigonometric functions. In other words, if θ is an angle with radian measure s, then any trigonometric function of θ is equal to the corresponding circular function of the real number s. Consequently, every identity in this text is valid for real numbers as well as for angles. Thus $\sin^2 s + \cos^2 s = 1$ holds true for every real number s.

EXERCISE 19

(Unless directed to the contrary, round off approximate results to three-figure accuracy.)

1. Find the number of radians in the central angle that intercepts an arc of 9 in on a circle of diameter 3 ft. Express the angle in degrees.

2. On a circle of radius 18.0 cm, how long is the arc intercepted by a central angle of (a) $\frac{2}{3}$ radian, (b) $80°$, (c) 1.7π radians?

3. Find the radius of a circle on which (a) a central angle of 0.7 radian intercepts an arc of 28 in, (b) a central angle of $32°$ intercepts an arc of 4.5 meters.

4. A highway curve is to be laid out on a circle. What radius should be used if the road is to change its direction by $29.1°$ in a distance of 100 meters?

5. The minute hand of the Big Ben clock in London is 14.0 ft long. How far does the tip of the hand travel in (a) 10 min, (b) 45 min, (c) 1 day?

6. The pendulum of a clock swings through an arc of precisely $8.0°$ each second. If the tip of the pendulum travels 831 cm in 1 min, how long is the pendulum?

In Problems 7 to 14, assume that the earth is a sphere of radius 4000 mi. Write results with two-figure accuracy.

7. How far is it from the equator to Kansas City, Mo. (latitude $39°$ N)? How far is Kansas City from the North Pole? (See Figure 43.)

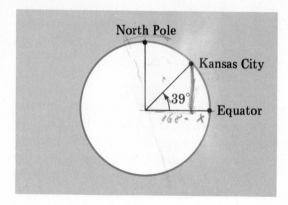

FIGURE 43

8. Edinburg, Tex., is 1800 mi from the equator. Find the latitude of Edinburg.

9. Dodge City, Kan., is 1470 mi due north of Acapulco, Mexico (latitude 17° N). Find the latitude of Dodge City.

10. Woonsocket, R.I. (latitude 42° N) is due north of Valparaiso, Chile (latitude 33° S). How far is Woonsocket from Valparaiso?

11. The North Pole is 480 mi closer to Bern, Switzerland, than to Pittsburgh, Pa. If the latitude of Bern is 47° N, find the latitude of Pittsburgh.

12. The latitude of University Park, Pa., is 41° N. The latitude of Melbourne, Australia, is 38° S. How many miles is University Park north of Melbourne?

13. Macapa, Brazil (longitude 51° W*) and Mbandaka, Zaire (longitude 18° E*) are located on the equator. Find the distance between the cities.

14. Entebbe, Uganda, and Sasak, Sumatra, are on the equator. Sasak is 4700 mi east of Entebbe (longitude 32.5° E). Find the longitude of Sasak.

In Problems 15 to 18, since the angle is small, we can assume, with little error, that the chord is equal to its subtended arc.

15. At a distance of 9610 ft, a television tower subtends an angle of 3.1° at the eye of an observer. Find the height of the tower. (See Figure 44.)

---

*51° W means 51° west of the prime meridian; 18° E means 18° east of it.

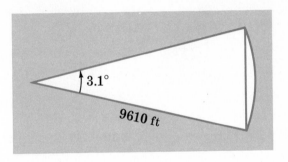

FIGURE 44

16. The diameter of the earth is 7930 mi. Find (to the nearest tenth of a degree) the angle subtended by the earth at the eye of an astronaut on the moon when our planet and its satellite are 241,000 mi apart.

17. A freight train standing on a straight north-south track subtends a horizontal angle of 2.5° at the eye of an observer who is due west of the middle of the train. If the length of the train is 0.110 km, how far is it from the observer?

18. The moon has a diameter of 2160 mi and is 240,000 mi from the earth. The sun has a diameter of 864,000 mi and is 93,000,000 mi from the earth. Compare the angle subtended by the moon at the earth with the angle subtended by the sun at the earth. Get results to the nearest thousandth of a degree.

19. On a circle of radius exactly 20, chords AB and AC are drawn. If the length of arc BAC is exactly 80, show that the acute angle BAC is $(\pi - 2)$ radians. *Hint:* An angle inscribed in a circle is measured by one-half its intercepted arc.

20. The radius of a circle is 100 cm. If A and B are two points on the circumference, and if the tangent lines at these points intersect at point T such that angle ATB is 36.0°, find the length of the minor arc AB.

21. A bucket is lifted from the bottom of a well by pulling a rope over a pulley. If the bucket is raised 102 cm by turning the pulley through 1.60 revolutions, what is the radius of the pulley?

22. The angle between a chord of a circle and the tangent line passing through one end of the chord is 1.2 radians. The length of the arc subtended by the chord is 36 cm. Find the radius of the circle.

23. A sector of a circle is the part of the circle bounded by an arc and the radii drawn to its extremities (Figure 45). By geometry, the area of a

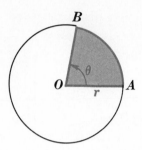

FIGURE 45

sector is equal to one-half its arc times the radius of the circle. Show that

$$\text{Area of sector } OAB = \tfrac{1}{2}r^2\theta \quad \text{where } \theta \text{ is in radians}$$

36 LINEAR AND ANGULAR VELOCITY

Consider a point P moving with constant speed on the circumference of a circle with radius r and center O. If P traverses a distance of s linear units (meters, feet, miles, etc.) in t time units (seconds, minutes, etc.), then $s/t = v$ is called the linear velocity of P. If the radius OP swings through θ angular units (degrees, radians, etc.), in t time units, then $\theta/t = \omega$ is called the angular velocity of P. If, further, θ is in radians and ω is in radians per unit of time, then we can divide $s = r\theta$ by t, and get

$$\frac{s}{t} = r \cdot \frac{\theta}{t} \qquad \frac{\theta}{t}$$

or

$$v = r\omega$$

provided ω is in radians per time unit. This means that the linear velocity of a point on the circumference of a circle is equal to the radius times the angular velocity of the point, in radians per unit of time.

The angular velocity of a rotating body is quite often expressed in revolutions per minute (rpm). This can be readily converted into radians per minute, by remembering that one revolution represents 2π radians.

EXAMPLE   A flywheel 6 ft in diameter makes 40 rpm. (*a*) Find its angular velocity in radians per second. (*b*) Find the speed of the belt that drives the flywheel.

SOLUTION   (*a*) Since 40 rpm represents $40(2\pi) = 80\pi$ radians per min,

$$\omega = \frac{80\pi}{60} = \frac{4\pi}{3} \text{ radians per sec}$$

(*b*) The speed of the belt, if it does not slip, is equal to the linear velocity of a point on the rim of the flywheel. Using $v = r\omega$, we have

$$v = 3\left(\frac{4\pi}{3}\right) = 4\pi \text{ ft/sec} \doteq 12.6 \text{ ft/sec}$$

EXERCISE 20

(*Unless directed to the contrary, round off approximate results to three-figure accuracy.*)

1. How many revolutions per minute are made by a wheel that has an angular velocity of 18,000 radians per hour?

2. A flywheel makes 1200 rpm. What is its angular velocity in radians per second?

3. A locomotive is traveling at 72.0 km/h. Find the diameter of the drive wheels in centimeters if they are making 250 rpm.

4. Find the linear speed, due to the rotation of the earth, of a point on the equator. Use 3960 mi as the radius at the equator. Express the result in miles per hour.

5. Find the linear speed, due to the rotation of the earth, of the Federal Reserve Building in Kansas City (latitude 39° N). Use 3960 mi as the radius of the earth. (See Figure 43.) Express the result in miles per hour.

6. The minute hand of the Big Ben clock is 14.0 ft long. Find the speed of the tip of the hand in inches per minute.

7. The tires of a bus have an outside diameter of 80.0 cm and are rotating at 500 rpm. How fast is the bus moving in kilometers per hour?

8. A truck is traveling at 120 km/h. How many revolutions per minute are made by the wheels, which are 80.0 cm in diameter?

9. The wheels of a car are rotating at 840 rpm. If the car is traveling at 55 mi/h, find the diameter of the wheels in inches.

10. The centers of two meshed gears are 20 cm apart. If the larger gear makes 2 rpm while the smaller one makes 3 rpm, find the radius of each gear.

11. A moving belt drives two flywheels with diameters of 60 cm and 80 cm. Find the speed of the belt in meters per second, and find the angular velocity of the smaller wheel in revolutions per minute if the larger wheel makes 72 rpm.

12. A belt moving 32 ft/sec is driving a pulley at the rate of 360 rpm. Find the diameter of the pulley in inches.

6 GRAPHS OF THE TRIGONOMETRIC FUNCTIONS

37 PERIODIC FUNCTIONS

Since coterminal angles have the same trigonometric functions, we know that any trigonometric function of θ is exactly equal to the same function of $\theta + n \cdot 360°$, where n is any integer. For example, $\sin 20° = \sin 380° = \sin 740° = \sin(-340°) = $ etc. Hence $\sin \theta$ takes on all of its possible values as θ moves from $0°$ to $360°$; in fact, it takes on all values between -1 and 1 *twice* for $0° \leq \theta < 360°$, once in increasing and once in decreasing, as θ increases (see Section 38). Then it repeats these values as θ moves from $360°$ to $720°$, from $720°$ to $1080°$, etc. Inasmuch as

$$\sin \theta = \sin(\theta + 360°) \qquad \text{for all } \theta$$

and since $360°$ is the smallest positive angle for which this is true, we say that the sine is a periodic function of period $360°$. In general,

> A function $f(\theta)$ is said to be periodic and of period p, provided
>
> $$f(\theta + p) = f(\theta) \qquad \text{for all } \theta$$
>
> where p is the smallest positive constant for which this is true.

The cosine function is periodic with a period of $360°$ because

$$\cos(\theta + 360°) = \cos \theta \qquad \text{for all } \theta$$

and because this statement is not true if $360°$ is replaced by any smaller positive angle, such as $180°$.

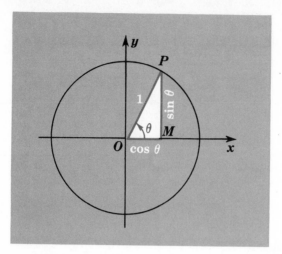

FIGURE 46

38 VARIATIONS OF THE SINE AND COSINE

Knowing that the sine and cosine functions have a period of 360°, we shall confine ourselves to investigating their behavior as the angle increases from 0° to 360°. Consider any angle θ in standard position (Figure 46). On the terminal side of θ, choose the point $P(x, y)$ whose radius vector is 1. As θ varies, make P move so that r is always 1. This will keep P on a circle with center O and radius 1. Recalling that $\sin \theta = y/r = MP/1 = MP$, we see that the length of the *directed* segment MP is equal to $\sin \theta$. Now try to visualize θ as increasing from 0° to 90°. Accordingly $\sin \theta$ (or MP) increases from 0 to 1. As θ increases from 90° to 180°, $\sin \theta$ decreases from 1 to 0. As θ swings through the third and fourth quadrants, P is below the x axis. Hence $\sin \theta$ (or MP) is negative.

Similarly, $\cos \theta = x/r = OM$. As θ increases from 0° to 90°, $\cos \theta$ decreases from 1 to 0. A study of the variations of $\sin \theta$ and $\cos \theta$ as θ goes from 0° to 360° reveals the following results:

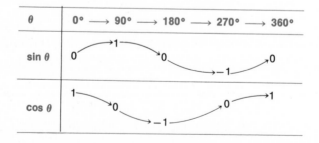

The arrow pointing upward means the function is increasing; downward, decreasing.

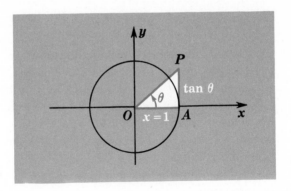

FIGURE 47

39 VARIATION OF THE TANGENT

In discussing $\tan \theta = y/x$, we shall keep x numerically equal to 1. For θ in Q I, we have $\tan \theta = AP$, the length of the tangent line to the unit circle at A (Figure 47). For $\theta = 0°$, P coincides with A, and $\tan 0° = 0$. As θ increases toward 90°, P moves upward from A, and $\tan \theta$ increases. When $\theta = 90°$, P is on the y axis and $x = 0$ (it cannot be 1). Hence $\tan 90° = y/0$, which *does not exist*. But as θ approaches 90°, $\tan \theta$ (or AP) increases rapidly. In fact we can make $\tan \theta$ just as large as we please by taking θ sufficiently close to 90°. A situation like this is expressed briefly by $\tan 90° = \infty$ (read "$\tan 90°$ is infinite"). Remember, however, that *$\tan 90° = \infty$ is just an abbreviation for: $\tan 90°$ does not exist; but by taking θ sufficiently close to 90° (never letting it equal 90°), we can make $\tan \theta$ as large numerically as we please.* Memorize this statement and think of it when you see the symbol ∞.

In Q II (Figure 48), keep $x = -1$; then P moves on the tangent line to the unit circle at B; and $\tan \theta = y/x = y/-1 = -y = -BP$ (a negative number since BP is positive). As soon as θ leaves 90°, P starts down on the

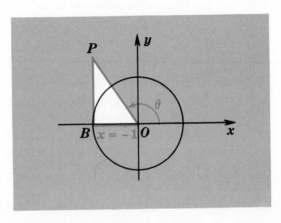

FIGURE 48

tangent line. As θ increases from 90° to 180°, BP decreases from very large positive numbers to 0, and $\tan \theta$ increases from very large negative values to 0. Using our symbolic notation, we say that as θ increases from 90° to 180°, $\tan \theta$ increases from $-\infty$ to 0.* By a similar process, the variation of $\tan \theta$ in Q III and Q IV can be investigated.

The variation of $\cot \theta$ can be studied by recalling that $\cot \theta = 1/\tan \theta$. Since $\tan \theta$ is always increasing, $\cot \theta$ is always decreasing. When $\tan \theta$ becomes infinite, $\cot \theta$ approaches 0; when $\tan \theta$ approaches 0, $\cot \theta$ becomes infinite.

The variations of $\sec \theta$ and $\csc \theta$ can be investigated through their reciprocals.

In summary we have

θ	Q I 0° ⟶ 90°	Q II 90° ⟶ 180°	Q III 180° ⟶ 270°	Q IV 270° ⟶ 360°
$\sin \theta$	0 ↗ 1	1 ↘ 0	0 ↘ −1	−1 ↗ 0
$\cos \theta$	1 ↘ 0	0 ↘ −1	−1 ↗ 0	0 ↗ 1
$\tan \theta$	0 ↗ ∞	−∞ ↗ 0	0 ↗ ∞	−∞ ↗ 0
$\cot \theta$	∞ ↘ 0	0 ↘ −∞	∞ ↘ 0	0 ↘ −∞
$\sec \theta$	1 ↗ ∞	−∞ ↗ −1	−1 ↘ −∞	∞ ↘ 1
$\csc \theta$	∞ ↘ 1	1 ↗ ∞	−∞ ↗ −1	−1 ↘ −∞

Notice that in Q I the sine, tangent, and secant increase while their cofunctions decrease. Also observe that the sine and cosine range from -1 to 1, the tangent and cotangent take on all values, and the secant and cosecant are always numerically equal to or greater than 1.

40 THE GRAPH OF sin θ

A complete "picture story" of the variation of the sine is presented by its graph. Let us draw a system of coordinate axes and label the horizontal

---

*The notation $\tan 90°^- = +\infty$ is frequently used to mean that as θ approaches 90° through values *less* than 90° (such as 89°, 89.9°, 89.99°, etc.), $\tan \theta$ *increases* without limit. The statement $\tan 90°^+ = -\infty$ is used to indicate that as θ approaches 90° through values *greater* than 90°, $\tan \theta$ *decreases* without limit. Both statements may be incorporated in the single form $\tan 90° = \infty$, which implies that as θ approaches 90° (from either side), $\tan \theta$ *increases numerically* without limit.

axis as θ and the vertical axis as $\sin \theta$. Since the radian is the natural measure of angles, let the θ axis be laid off in radians. This means that the number 1 on the vertical scale and the number 1 (radian) on the horizontal scale are represented by the same distance. *Hence 180° on the θ axis should be π times as long as 1 unit on the ($\sin \theta$) axis.* Using trigonometric tables and the related-angle theory, we form the following table:

θ in degrees	0°	30°	60°	90°	120°	150°	180°	210°	240°	270°	300°	330°	360°
θ in radians	0	$\dfrac{\pi}{6}$	$\dfrac{\pi}{3}$	$\dfrac{\pi}{2}$	$\dfrac{2\pi}{3}$	$\dfrac{5\pi}{6}$	π	$\dfrac{7\pi}{6}$	$\dfrac{4\pi}{3}$	$\dfrac{3\pi}{2}$	$\dfrac{5\pi}{3}$	$\dfrac{11\pi}{6}$	2π
$\sin \theta$	0	0.5	0.87	1	0.87	0.5	0	−0.5	−0.87	−1	−0.87	−0.5	0

After plotting these values on the coordinate axes,* we obtain the curve in Figure 49.

Students should practice drawing the 0° to 360° portion of this curve until they can make a hasty sketch of it from memory. The sine curve† can be used to remember the sine of 0°, 90°, 180°, and 270° and also to remember the sign of the sine in the various quadrants. For example, if you encounter $\sin 270°$, make a rapid sketch or form a mental picture of the sine curve and notice that $\sin 270° = -1$. Also, the sine curve

---

*A very good approximation to the sine curve can be drawn on a special kind of graph paper called quadrille paper. Let 4 quadrille *spaces* represent 1 *unit* vertically and let 1 quadrille space represent 15° horizontally. For $\theta = 15°$, $\sin \theta = 0.26$; the corresponding point would be plotted 1 *space* to the right and $4(0.26) = 1.04$ *spaces* up from the origin. For $\theta = 45°$, $\sin \theta = 0.71$; the corresponding point would be plotted 3 *spaces* to the right and $4(0.71) = 2.88$ *spaces* up from the origin.

† Sometimes called the *sinusoid.*

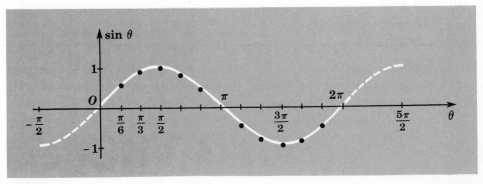

FIGURE 49

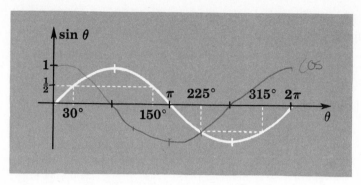

FIGURE 50

enables us to remember that sin θ is positive in Q I and Q II (because in these quadrants the curve is above the horizontal axis) and negative in Q III and Q IV (because the curve is below the horizontal axis for $180° < \theta < 360°$).

Furthermore, the sine curve recalls the related-angle theory. For example, it is obvious from the curve (see Figure 50) that

$$\sin 150° = \sin 30° = \tfrac{1}{2}$$

and

$$\sin 225° = \sin 315° = -\frac{\sqrt{2}}{2}$$

Because of its wave form, the sine curve is very important in the study of wave motion in electrical engineering and physics. The maximum distance of the curve from the θ axis is called the *amplitude* of the curve or wave. The period of the function representing the wave is called the period (or *wavelength*) of the curve. The sine curve has an amplitude of 1 and a period of 2π. The student can see from Figure 50 that sin θ does not have a period that is less than $360°$.

41 GRAPHS OF THE OTHER TRIGONOMETRIC FUNCTIONS

By using methods exactly like those employed in the preceding section, we can draw the graphs of the other trigonometric functions. For reference we exhibit these graphs (Figures 51 to 55). Students should draw each of them by preparing a table of values, plotting the points, and then drawing a smooth curve through these points. From these graphs we can see that tan θ and cot θ have a period of π, whereas sec θ and csc θ have a period 2π.

sin 30   = cos 60

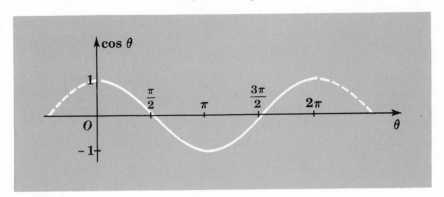

FIGURE 51

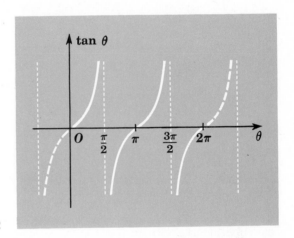

FIGURE 52

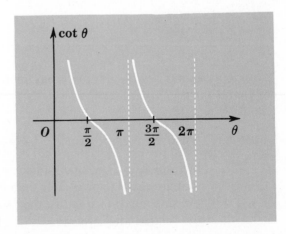

FIGURE 53

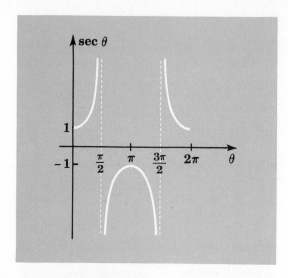

FIGURE 54

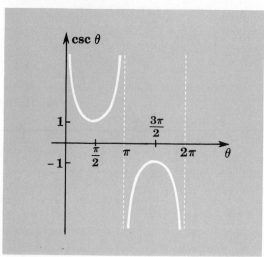

FIGURE 55

Students should be able to draw from memory a hasty sketch of $\sin \theta$, $\cos \theta$, and $\tan \theta$.

EXAMPLE   Solve the equation $\sin \theta = -\dfrac{\sqrt{3}}{2}$ for all values of θ on the interval $0° \leq \theta < 360°$. (That is, solve for all values of θ that are equal to or greater than $0°$ and less than $360°$.) Use the sine curve to identify the proper quadrants and check the related-angle theory.

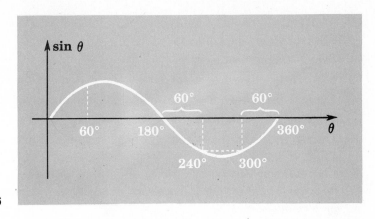

FIGURE 56

SOLUTION   The sine is negative (the sine curve lies below the θ axis) in Q III and Q IV. Remember that $\sin 60° = \dfrac{\sqrt{3}}{2}$. Since $\sin \theta = -\dfrac{\sqrt{3}}{2}$, it follows that θ must be equal to those angles in Q III and Q IV that have 60° for their related angle (see Figure 56).

In Q III: $\theta = 180° + 60° = 240°$
In Q IV: $\theta = 360° - 60° = 300°$

Hence if

$$\sin \theta = -\frac{\sqrt{3}}{2}$$

then $\theta = 240°, 300°$

EXERCISE 21

1. Sketch the sine curve by drawing a smooth curve through the points obtained by assigning to θ the values $-90°$, $-75°$, $-60°$, $-45°$, $-30°$, $-15°$, $0°$, $15°$, $30°$, ..., $360°$.
2. Sketch the cosine curve. Locate points every 15° from 0° to 360°.
3. Sketch the tangent curve. Locate points every 15° from $-90°$ to 360°.

4. Sketch the cotangent curve.

5. Sketch the secant curve.

6. Sketch the cosecant curve.

7. Explain briefly what is meant by the statement "sec 270° = ∞."

8. Explain briefly what is meant by the statement "cot 180° = ∞."

9. What is the period of tan θ?

10. Discuss in detail the variation of the secant.

Solve the following equations for values of θ on the interval $0° \leq \theta < 360°$. Use the curves to identify the proper quadrants and check the related-angle theory. Do not use a calculator or tables.

11. $\sin \theta = \dfrac{\sqrt{3}}{2}$
    $\qquad$ 12. $\cos \theta = \dfrac{\sqrt{3}}{2}$

13. $\cos \theta = -\dfrac{\sqrt{3}}{2}$
    $\qquad$ 14. $\sin \theta = -\frac{1}{2}$

15. $\cos \theta = -\dfrac{\sqrt{2}}{2}$
    $\qquad$ 16. $\sin \theta = -1$

17. $\sin \theta = \dfrac{\sqrt{2}}{2}$
    $\qquad$ 18. $\cos \theta = 0$

19. $\csc \theta = 2$
    $\qquad$ 20. $\cot \theta = 1$

21. $\tan \theta = -\sqrt{3}$
    $\qquad$ 22. $\sec \theta = \sqrt{2}$

Solve the following equations for values of θ on the interval $0 \leq \theta < 2\pi$ (radians). Do not use a calculator or tables.

23. $\cos \theta = -1$
    $\qquad$ 24. $\cos \theta = -\frac{1}{2}$

25. $\sin \theta = 0$
    $\qquad$ 26. $\sec \theta = 2$

27. $\cot \theta = \dfrac{1}{\sqrt{3}}$
    $\qquad$ 28. $\csc \theta = \sqrt{2}$

29. $\sec \theta = 0$
    $\qquad$ 30. $\tan \theta = -1$

Use a calculator or tables to solve the following equations for values of θ on the interval $0° \leq \theta < 360°$. Obtain results correct to the nearest degree.

31. $\cos \theta = -\frac{2}{3}$
    $\qquad$ 32. $\sin \theta = \frac{1}{6}$

33. $\cot \theta = 100$
    $\qquad$ 34. $\tan \theta = -10$

35. $\tan \theta = 0.8$
    $\qquad$ 36. $\cot \theta = -0.07$

37. $\sin \theta = -0.2443$
    $\qquad$ 38. $\cos \theta = 0.5299$

39. State the maximum and minimum values achieved by each of the following functions of θ:
   (a) $20 + \sin 2\theta$
   (b) $4 - 7 \cos \theta$
   (c) $-5 + \sin^2 \theta$

40. State the maximum and minimum values achieved by each of the following functions of θ:
   (a) $15 + 6 \cos 2\theta$
   (b) $9 - 8 \sin \theta$
   (c) $21 - \cos^4 \theta$

41. Correct the following statement: $\cot 180° = \infty$ means (a) $\cot 180°$ does not exist, and (b) by taking θ sufficiently close to $180°$, we can make $\cot 180°$ as large numerically as we please.

In Problems 42 to 46, either prove or disprove the statement. (Prove that the equation is, or is not, an identity.)

42. $\sin n\pi = 0$ (n an integer)

43. $\cos n\pi = (-1)^n$ (n an integer)

44. $\cos \dfrac{n\pi}{2} = 0$ (n an odd integer)

45. $\sin \dfrac{n\pi}{2} = (-1)^{(n-1)/2}$ (n an odd integer)

46. $\tan \dfrac{n\pi}{4} = (-1)^{(n-1)/2}$ (n an odd integer)

7 FUNCTIONS OF TWO ANGLES

42 FUNCTIONS OF THE SUM OF TWO ANGLES

The trigonometric identities in Chapter 3 are relations among the trigonometric functions of one angle. We shall now consider functions of an angle which is the sum of two given angles. It seems reasonable to say that if the functions of 30° and 45° are known, the functions of 75° can be obtained. For instance, is sin 30° + sin 45° = sin 75°? Obviously this is false, because $\dfrac{1}{2} + \dfrac{\sqrt{2}}{2} = 0.5 + 0.7 = 1.2$, which would make sin 75° greater than 1. This proves that, in general, sin $(A + B) \neq$ sin A + sin B.* Likewise sin $2A$ is not identically equal to 2 sin A because sin 60° $\neq$ 2 sin 30°. It is, however, possible to express sin $(A + B)$ in terms of the functions of the separate angles A and B. And it is possible to express sin $2A$ in terms of the functions of A. These and other formulas will be developed in the following sections.

43 sin $(A + B)$ AND cos $(A + B)$†

The purpose of this section is to prove the following identities:

$$\sin (A + B) = \sin A \cos B + \cos A \sin B \qquad (1)$$
$$\cos (A + B) = \cos A \cos B - \sin A \sin B \qquad (2)$$

---

*The symbol $\neq$ is read "is not equal to."

†The instructor may wish to begin with a simple geometric proof (Section 84).

106

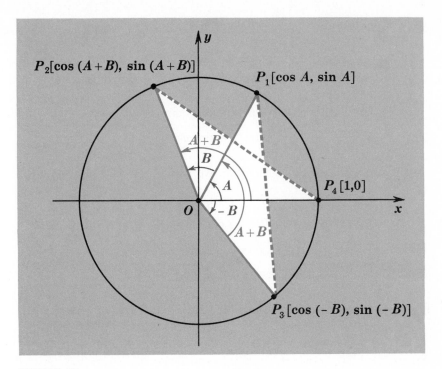

FIGURE 57

We shall first derive a formula for cos $(A + B)$. On a coordinate system (Figure 57), draw a unit circle with center at the origin. Place the angles A, $A + B$, and $-B$ in standard position. Let the terminal sides of these angles intersect the unit circle at points P_1, P_2, and P_3, respectively; let P_4 designate the point $(1, 0)$. If P_1 has coordinates (x_1, y_1), then cos $A = x_1/1 = x_1$, and sin $A = y_1/1 = y_1$. Hence the coordinates of P_1 are (cos A, sin A).* For the same reason, the coordinates of P_2 and P_3 are as indicated in Figure 57. Angle P_3OP_1 is $(B + A)$ because angle P_3OP_4 is B and angle P_4OP_1 is A. Therefore, triangles P_3OP_1 and P_4OP_2 are congruent. Why? Hence $P_3P_1 = P_4P_2$.

Applying the distance formula (Section 4), we get

$$P_3P_1 = \sqrt{[\cos A - \cos (-B)]^2 + [\sin A - \sin (-B)]^2}$$

Squaring and using the identities cos $(-B) = \cos B$ and sin $(-B) =$

---

*This statement is true regardless of the quadrant in which A lies.

$-\sin B$, we obtain

$$
\begin{aligned}
(P_3P_1)^2 &= (\cos A - \cos B)^2 + (\sin A + \sin B)^2 \\
&= \cos^2 A - 2 \cos A \cos B + \cos^2 B + \sin^2 A + 2 \sin A \sin B \\
&\quad + \sin^2 B \\
&= 2 - 2 \cos A \cos B + 2 \sin A \sin B
\end{aligned}
$$

since $$\cos^2 \theta + \sin^2 \theta = 1$$

Again using the distance formula, we find

$$
\begin{aligned}
(P_4P_2)^2 &= [\cos (A + B) - 1]^2 + [\sin (A + B) - 0]^2 \\
&= \cos^2 (A + B) - 2 \cos (A + B) + 1 + \sin^2 (A + B)
\end{aligned}
$$

Hence $$(P_4P_2)^2 = 2 - 2 \cos (A + B)$$

since $$\cos^2 (A + B) + \sin^2 (A + B) = 1$$

Upon equating the expressions for $(P_4P_2)^2$ and $(P_3P_1)^2$, we get

$$2 - 2 \cos (A + B) = 2 - 2 \cos A \cos B + 2 \sin A \sin B$$

which reduces to

$$\cos (A + B) = \cos A \cos B - \sin A \sin B \tag{2}$$

Note carefully that this proof does not depend upon the quadrants in which A, B, and $A + B$ happen to lie. The formula for $\cos (A + B)$ is valid for all values of A and B, positive or negative.

Before deriving a formula for $\sin (A + B)$, we shall need to obtain general expressions for $\cos (A - C)$, $\cos (90° - C)$, and $\sin (90° - C)$.

Since formula (2) for $\cos (A + B)$ holds for all A and B, we shall rewrite it with B replaced by $-C$. Hence

$$
\begin{aligned}
\cos (A - C) &= \cos [A + (-C)] \\
&= \cos A \cos (-C) - \sin A \sin (-C)
\end{aligned}
$$

Since $\cos (-C) = \cos C$ and $\sin (-C) = -\sin C$, we have

$$\cos (A - C) = \cos A \cos C + \sin A \sin C \tag{7-1}$$

If we replace A with $90°$ in formula (7-1), we find

$$\cos (90° - C) = \cos 90° \cos C + \sin 90° \sin C$$

Since $\cos 90° = 0$ and $\sin 90° = 1$, we have

$$\cos (90° - C) = \sin C \qquad\qquad (7\text{-}2)$$

Moreover, since $C = 90° - (90° - C)$,

$$\begin{aligned}
\cos C &= \cos [90° - (90° - C)] \\
&= \cos 90° \cos (90° - C) + \sin 90° \sin (90° - C) \\
&= 0 \cdot \cos (90° - C) + 1 \cdot \sin (90° - C)
\end{aligned}$$

Hence $\cos C = \sin (90° - C)$

or $$\sin (90° - C) = \cos C \qquad\qquad (7\text{-}3)$$

Formulas (7-2) and (7-3) were established for any acute angle C in Section 11. The foregoing arguments prove the formulas for *all* values of C. For example, if $C = 250°$, formula (7-2) says $\cos (90° - 250°) = \sin 250°$, which is true because $\cos (90° - 250°) = \cos (-160°) = \cos 160° = -\cos 20° = -\sin 70°$, whereas $\sin 250° = -\sin 70°$, by the related-angle theorem (Section 29).

To derive a formula for $\sin (A + B)$, we shall reverse (7-2) and replace C with $(A + B)$ to get

$$\begin{aligned}
\sin (A + B) &= \cos [90° - (A + B)] \\
&= \cos [(90° - A) - B] \\
&= \cos (90° - A) \cos B + \sin (90° - A) \sin B
\end{aligned}$$

Using formulas (7-2) and (7-3), we may write

$$\sin (A + B) = \sin A \cos B + \cos A \sin B \qquad\qquad (1)$$

It is impossible to overemphasize the importance of these results. In addition to learning formulas (1) and (2), the student should be able to state them in words. These statements are

(1) The sine of the sum of two angles is equal to the sine of the first times the cosine of the second, plus the cosine of the first times the sine of the second.

(2) The cosine of the sum of two angles is equal to the cosine of the first times the cosine of the second, minus the sine of the first times the sine of the second.

It is equally important for the student to be able to use these formulas backward. For example, when

$$\cos 7\theta \cos 2\theta - \sin 7\theta \sin 2\theta$$

is encountered, it should be recognized as the expansion of $\cos (7\theta + 2\theta)$ or $\cos 9\theta$.

EXAMPLE 1 Compute $\sin 75°$ and $\cos 75°$ from the functions of $30°$ and $45°$.

SOLUTION

$$\sin 75° = \sin (30° + 45°)$$

$$= \sin 30° \cos 45° + \cos 30° \sin 45°$$

$$= \frac{1}{2} \cdot \frac{\sqrt{2}}{2} + \frac{\sqrt{3}}{2} \cdot \frac{\sqrt{2}}{2}$$

$$= \frac{\sqrt{2}}{4} + \frac{\sqrt{6}}{4} = \frac{\sqrt{2} + \sqrt{6}}{4}$$

$$\doteq \frac{1.414 + 2.449}{4} = \frac{3.863}{4} \doteq 0.966$$

$$\cos 75° = \cos (30° + 45°)$$

$$= \cos 30° \cos 45° - \sin 30° \sin 45°$$

$$= \frac{\sqrt{3}}{2} \cdot \frac{\sqrt{2}}{2} - \frac{1}{2} \cdot \frac{\sqrt{2}}{2}$$

$$= \frac{\sqrt{6} - \sqrt{2}}{4} \doteq 0.259$$

Notice that these decimal approximations agree with the "story" as presented by the sine and cosine curves. Angles near $90°$ have sines close to 1 and cosines close to 0.

EXAMPLE 2 Given $\sin A = \frac{5}{13}$, with A in Q I, and $\cos B = -\frac{4}{5}$, with B in Q II. Find $(a) \sin (A + B)$, $(b) \cos (A + B)$, (c) the quadrant in which $(A + B)$ lies.

SOLUTION First find $\cos A$ and $\sin B$ by drawing triangles of reference (Figure 58). Hence

$$\cos A = \tfrac{12}{13} \quad \text{and} \quad \sin B = \tfrac{3}{5}$$

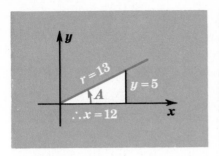 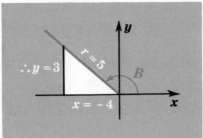

FIGURE 58

(a) $\sin (A + B) = \sin A \cos B + \cos A \sin B$

$$= \left(\frac{5}{13}\right)\left(\frac{-4}{5}\right) + \left(\frac{12}{13}\right)\left(\frac{3}{5}\right)$$

$$= -\tfrac{20}{65} + \tfrac{36}{65} = \tfrac{16}{65}$$

(b) $\cos (A + B) = \cos A \cos B - \sin A \sin B$

$$= \left(\frac{12}{13}\right)\left(\frac{-4}{5}\right) - \left(\frac{5}{13}\right)\left(\frac{3}{5}\right)$$

$$= -\tfrac{48}{65} - \tfrac{15}{65} = -\tfrac{63}{65}$$

(c) The angle $(A + B)$ lies in Q II because $\sin (A + B)$ is positive and $\cos (A + B)$ is negative.

EXERCISE 22

1. Compute $\sin 195°$ from the functions of $60°$ and $135°$.
2. Compute $\cos 345°$ from the functions of $30°$ and $315°$.
3. Compute $\sin 285°$ from the functions of $240°$ and $45°$.
4. Compute $\cos 165°$ from the functions of $45°$ and $120°$.
5. Use a calculator or tables to show that

$$\sin 40° + \sin 50° \neq \sin 90°$$
$$\cos 20° + \cos 35° \neq \cos 55°$$

Simplify by reducing to a single term.

6. $\cos 200° \cos 160° - \sin 200° \sin 160°$

7. $\sin \theta \sin 2\theta - \cos \theta \cos 2\theta$

8. $\sin \dfrac{\theta}{6} \cos \dfrac{5\theta}{6} + \cos \dfrac{\theta}{6} \sin \dfrac{5\theta}{6}$

9. $\sin 80° \cos 100° + \sin 100° \cos 80°$

Prove the following identities:

10. $\sin (\pi + \theta) = -\sin \theta$

11. $\sin \left(\dfrac{7\pi}{2} + 4\theta \right) = -\cos 4\theta$

12. $\cos \left(-\dfrac{\pi}{2} + 5\theta \right) = \sin 5\theta$

13. $\cos (C - D) \cos D - \sin (C - D) \sin D = \cos C$
  Hint: $C = (C - D) + D$.

14. $\sin (45° + \theta) - \cos (225° + \theta) = \sqrt{2} \cos \theta$

15. $\sin (30° + \theta) + \cos (120° + \theta) = 0$

16. Express $\sin 49°$ in terms of functions of $40°$ and $9°$.

17. Express $\cos 77°$ in terms of functions of $11°$ and $66°$.

18. Prove $\quad \dfrac{\sin \theta}{\sin 5\theta} + \dfrac{\cos \theta}{\cos 5\theta} = \dfrac{\sin 6\theta}{\sin 5\theta \cos 5\theta}$.

19. Prove $\quad \dfrac{\cos 7\theta}{\sin 2\theta} - \dfrac{\sin 7\theta}{\cos 2\theta} = \cos 9\theta \sec 2\theta \csc 2\theta$.

20. Given $\sin A = \frac{1}{10}$ with A in Q I, and $\sin B = \frac{9}{10}$ with B in Q I. Find $\sin (A + B)$. Check your result by using a calculator or tables to approximate A, B, and $(A + B)$ as determined by their sines.

21. Given $\cos A = \frac{8}{17}$ with A in Q I, and $\tan B = -\frac{3}{4}$ with B in Q II. Find (a) $\sin (A + B)$, (b) $\cos (A + B)$, (c) the quadrant in which $(A + B)$ lies.

22. Given $\sin A = \frac{4}{5}$ with A in Q I, and $\cos B = -\frac{7}{25}$ with B in Q III. Find (a) $\sin (A + B)$, (b) $\cos (A + B)$, (c) the quadrant in which $(A + B)$ lies.

23. Given $\sin (C - D) = \frac{40}{41}$ with $(C - D)$ in Q I, and $\sin D = \frac{5}{13}$ with D in Q I. Find $\sin C$.

24. Given $\cot A = \frac{12}{5}$ with A in Q I, and $\cos B = -\frac{15}{17}$ with B in Q III. Find (a) $\sin (A + B)$, (b) $\cos (A + B)$, (c) the quadrant in which $(A + B)$ lies.

25. Prove the identity

$$\sin (A + B + C) = \sin A \cos B \cos C + \cos A \sin B \cos C$$
$$+ \cos A \cos B \sin C - \sin A \sin B \sin C$$

Hint: $\sin (A + B + C) = \sin [(A + B) + C]$.

26. Prove the identity

$$\cos (A + B + C) = \cos A \cos B \cos C - \sin A \sin B \cos C$$
$$- \sin A \cos B \sin C - \cos A \sin B \sin C$$

27. If A and B are complementary angles, prove that

$$\cos (2A + 9B) = -\cos 7B$$

28. If A and B are complementary angles, prove that

$$\sin (4A + B) = \cos 3A$$

29. Prove

$$(\cos A \cos B - \sin A \sin B)^2 + (\cos A \sin B + \sin A \cos B)^2 = 1$$

In Problems 30 to 32, either prove or disprove the statement. (Prove that the equation is or is not an identity.)

30. $\cos (\theta + n\pi) = (-1)^n \cos \theta$ (n an integer)

31. $\sin \left(\theta + \dfrac{n\pi}{2}\right) = (-1)^{(n-1)/2} \cos \theta$ (n an odd integer)

32. $\cos \left(\theta + \dfrac{n\pi}{2}\right) = (-1)^{(n+1)/2} \sin \theta$ (n an odd integer)

44 tan (A + B)

Since

$$\tan \theta = \frac{\sin \theta}{\cos \theta}$$

$$\tan (A + B) = \frac{\sin (A + B)}{\cos (A + B)} = \frac{\sin A \cos B + \cos A \sin B}{\cos A \cos B - \sin A \sin B}$$

Dividing top and bottom of this fraction by $\cos A \cos B$, we get

$$\tan (A + B) = \frac{\dfrac{\sin A \cos B}{\cos A \cos B} + \dfrac{\cos A \sin B}{\cos A \cos B}}{\dfrac{\cos A \cos B}{\cos A \cos B} - \dfrac{\sin A \sin B}{\cos A \cos B}}$$

$$= \frac{\dfrac{\sin A}{\cos A} + \dfrac{\sin B}{\cos B}}{1 - \dfrac{\sin A}{\cos A} \cdot \dfrac{\sin B}{\cos B}}$$

or
$$\tan (A + B) = \frac{\tan A + \tan B}{1 - \tan A \tan B} \tag{3}$$

It is important to note that this formula, and all others developed in this chapter, is valid for only those angles for which the functions involved are defined and for which the denominators are not 0.

Stated in words, we have

> (3) The tangent of the sum of two angles is equal to the sum of their tangents, divided by 1 minus the product of their tangents.

45 sin (A − B), cos (A − B), AND tan (A − B)

If A and B are any two angles, then,

$$\sin (A - B) = \sin A \cos B - \cos A \sin B \tag{4}$$

$$\cos (A - B) = \cos A \cos B + \sin A \sin B \tag{5}$$

$$\tan (A - B) = \frac{\tan A - \tan B}{1 + \tan A \tan B} \tag{6}$$

PROOF   Recall that $\sin (-B) = -\sin B$, $\cos (-B) = \cos B$, and $\tan (-B) = -\tan B$. Then

$$
\begin{aligned}
\sin (A - B) &= \sin [A + (-B)]^* \\
&= \sin A \cos (-B) + \cos A \sin (-B) \\
&= \sin A \cos B + (\cos A)(-\sin B) \\
&= \sin A \cos B - \cos A \sin B
\end{aligned}
$$

---

*This method may be used to convert any "sum formula" to a "difference formula."

Formulas (5) and (6) are proved by a similar method. The student should state formulas (4), (5), and (6) in words.

In comparing (1), (4), (2), and (5),

$$\sin (A + B) = \sin A \cos B + \cos A \sin B$$
$$\sin (A - B) = \sin A \cos B - \cos A \sin B$$
$$\cos (A + B) = \cos A \cos B - \sin A \sin B$$
$$\cos (A - B) = \cos A \cos B + \sin A \sin B$$

we notice that "the sines have the same sign, but the cosines have different signs."

There is no need for formulas involving the cotangent, secant, and cosecant of $(A + B)$ and $(A - B)$ because they can be readily expressed in terms of their reciprocals.

46 REDUCTION OF   *a* sin θ + *b* cos θ   TO   *k* sin (θ + *H*)

Any two real nonzero numbers a and b are proportional to two other numbers that represent the cosine and sine, respectively, of some properly chosen angle H. These other two numbers are $a/\sqrt{a^2 + b^2}$ and $b/\sqrt{a^2 + b^2}$. We may write

$$\cos H = \frac{a}{\sqrt{a^2 + b^2}} \quad \text{and} \quad \sin H = \frac{b}{\sqrt{a^2 + b^2}} \quad (7\text{-}4)$$

as indicated by Figure 59. Then

$$a \sin \theta + b \cos \theta = \sqrt{a^2 + b^2} \left[(\sin \theta) \frac{a}{\sqrt{a^2 + b^2}} + (\cos \theta) \frac{b}{\sqrt{a^2 + b^2}} \right]$$

$$= \sqrt{a^2 + b^2} \,(\sin \theta \cos H + \cos \theta \sin H)$$

Hence $a \sin \theta + b \cos \theta = \sqrt{a^2 + b^2} \sin (\theta + H)$

where H is an angle that satisfies both equations in (7-4). *

---

*Equations (7-4) imply that $\tan H = b/a$. In many cases it will be easiest to obtain H through its tangent, though then we must be careful to choose an angle in the correct quadrant as determined by the signs of a and b.

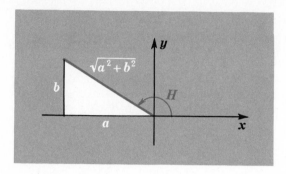

FIGURE 59

EXAMPLE   Express $4 \sin \theta + 3 \cos \theta$ in the form $k \sin (\theta + H)$, where k and H are constants and $k > 0$.

SOLUTION   Multiply and divide the given expression by $\sqrt{4^2 + 3^2} = 5$ to get

$$5(\tfrac{4}{5} \sin \theta + \tfrac{3}{5} \cos \theta) \qquad \text{or} \qquad 5[(\sin \theta)\tfrac{4}{5} + (\cos \theta)\tfrac{3}{5}]$$

Put $\tfrac{4}{5} = \cos H$; then $\tfrac{3}{5} = \sin H$. Our expression becomes

$$5(\sin \theta \cos H + \cos \theta \sin H) = 5 \sin (\theta + H)$$

Using a calculator or tables, we find that if $\cos H = \tfrac{4}{5}$ and $\sin H > 0$, then $H = 36.87°$ or 0.6435 radian. Hence

$$4 \sin \theta + 3 \cos \theta = 5 \sin (\theta + 36.87°)$$
$$= 5 \sin (\theta + 0.64) \qquad \text{approx.}$$

EXERCISE 23

Do not use a calculator or tables in Problems 1 to 39.

1. Compute $\cos 285°$ from the functions of $315°$ and $30°$.

2. Compute $\sin 75°$ from the functions of $135°$ and $60°$.

3. Compute $\tan 105°$ from $\tan 150°$ and $\tan 45°$.

4. Compute $\tan 255°$ from $\tan 30°$ and $\tan 225°$.

Simplify by reducing to a single term.

5. $\dfrac{\tan 85° - \tan 25°}{1 + \tan 85° \tan 25°}$

6. $\dfrac{\tan 7\theta + \tan 2\theta}{1 - \tan 7\theta \tan 2\theta}$

7. $\cos 6\theta \cos 4\theta + \sin 6\theta \sin 4\theta$

8. $\sin 77° \cos 17° - \cos 77° \sin 17°$

Prove the following identities.

9. $\tan \left(\theta + \dfrac{\pi}{6} \right) = \dfrac{1 + \sqrt{3} \tan \theta}{\sqrt{3} - \tan \theta}$ **10.** $\tan \left(\theta - \dfrac{\pi}{4} \right) = \dfrac{\tan \theta - 1}{\tan \theta + 1}$

11. $\sin (\pi - \theta) = \sin \theta$ **12.** $\cos (\theta - 270°) = -\sin \theta$

13. Given $\sin A = -\frac{24}{25}$ with A in Q III, and $\cos B = \frac{12}{13}$ with B in Q I. Find   (a) $\sin (A - B)$,   (b) $\cos (A - B)$,   (c) $\tan (A + B)$, (d) $\tan (A - B)$.

14. Given $\cos A = \frac{15}{17}$ with A in Q IV, and $\tan B = \frac{40}{9}$ with B in Q I. Find (a) $\sin (A - B)$, (b) $\cos (A - B)$, (c) $\tan (A + B)$, (d) $\tan (A - B)$.

15. Given $\tan A = \frac{1}{3}$ with A in Q I, and $\tan B = \frac{1}{2}$ with B in Q I. Prove $A + B = 45°$.

16. Given $\tan (C + D) = 3$ with $(C + D)$ in Q I, and $\tan D = 2$. Find $\tan C$.

Prove or disprove. (For each of the following equations, show that it is or is not an identity.)

17. $\tan \theta + \tan 180° = \tan (\theta + 180°)$

18. $\cos (A - B) + \cos B = \cos A$

19. $\sin^2 (A - B) = 1 - \cos^2 (B - A)$

20. $8 \cos 5\theta \cos 3\theta + 8 \sin 5\theta \sin 3\theta = 16 \cos 2\theta$

21. Express $\cos 23°$ in terms of functions of $79°$ and $56°$.

22. Express $\sin 61°$ in terms of functions of $27°$ and $88°$.

23. Express $\tan 12°$ in terms of $\tan 36°$ and $\tan 24°$.

24. Express $\tan 88°$ in terms of $\tan 33°$ and $\tan 55°$.

Prove the following identities:

25. $\csc (A - B) = \dfrac{\csc A \csc B}{\cot B - \cot A}$

26. $\dfrac{\sin 6\theta}{\csc 4\theta} + \dfrac{\cos 6\theta}{\sec 4\theta} = \cos 2\theta$

27. $\cos \left(\theta - \dfrac{\pi}{4} \right) - \sin \left(\dfrac{5\pi}{4} - \theta \right) = \sqrt{2} \cos \theta$

28. $\sin (330° - \theta) - \cos (240° - \theta) = 0$

29. $\cot (A - B) = \dfrac{\cot A \cot B + 1}{\cot B - \cot A}$

30. $\tan (A + B + C) = \dfrac{\tan A + \tan B + \tan C - \tan A \tan B \tan C}{1 - \tan A \tan B - \tan A \tan C - \tan B \tan C}$

31. $\cot (A + B) = \dfrac{\cot A \cot B - 1}{\cot A + \cot B}$

32. $\dfrac{\tan (A + B) - \tan C}{1 + \tan (A + B) \tan C} = \dfrac{\tan A + \tan (B - C)}{1 - \tan A \tan (B - C)}$

33. $\sin (\pi - C - D) = \sin (C + D)$

34. In the accompanying figure, if $a > b$, show that $x = b \sqrt{\dfrac{a + b}{a - b}}$.

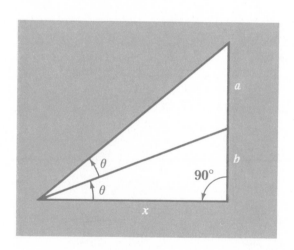

35. Prove $\sin (A - B) \sin (A + B) = \cos^2 B - \cos^2 A$.

36. If A and B are complementary angles, prove that $\sin (4A - 3B) = \cos 7A$.

37. If A and B are complementary angles, prove that $\cos 2A = -\cos 2B$.

38. If A and B are supplementary angles and n is an integer, prove that $\tan (A - nB) = -\tan (n + 1)B$.

39. If n is an integer, prove that $\sin (n\pi - \theta) = (-1)^{n+1} \sin \theta$.

40. Express $3 \sin \theta + 4 \cos \theta$ in the form $k \sin (\theta + H)$.

41. Express $\sin \theta - \cos \theta$ in the form $k \sin (\theta + H)$.

42. Express $-12 \sin \theta + 5 \cos \theta$ in the form $k \sin (\theta + H)$.

43. Express $24 \sin \theta + 7 \cos \theta$ in the form $k \sin (\theta + H)$.

44. Express $8 \sin \left(\theta + \dfrac{3\pi}{4} \right)$ in the form $a \sin \theta + b \cos \theta$.

45. Express $100 \sin \left(\theta + \dfrac{7\pi}{6} \right)$ in the form $a \sin \theta + b \cos \theta$.

47 DOUBLE–ANGLE FORMULAS

If A is any angle, then

$$\sin 2A = 2 \sin A \cos A \tag{7}$$

$$\cos 2A = \cos^2 A - \sin^2 A \tag{8a}$$

$$= 1 - 2 \sin^2 A \tag{8b}$$

$$= 2 \cos^2 A - 1 \tag{8c}$$

PROOF

$$\sin 2A = \sin (A + A)$$
$$= \sin A \cos A + \cos A \sin A$$
$$= 2 \sin A \cos A$$

$$\cos 2A = \cos (A + A)$$
$$= \cos A \cos A - \sin A \sin A$$
$$= \cos^2 A - \sin^2 A$$
$$= 1 - \sin^2 A - \sin^2 A$$
$$= 1 - 2 \sin^2 A \qquad \text{etc.}$$

Stated in words, formula (7) says

> The sine of twice an angle is equal to twice the sine of the angle times the cosine of the angle.

The student should state the three forms of formula (8) in words. Formulas (7) and (8) are called *double-angle formulas* because the angle on the left side is double the angle on the right side. Formula (7) could have been written

$$\sin B = 2 \sin \frac{B}{2} \cos \frac{B}{2}$$

because B is the double of $B/2$. In other words, formulas (7) and (8) are used to express the sine and cosine of an angle in terms of the functions of an angle that is half as large. To illustrate,

$$\sin 60° = 2 \sin 30° \cos 30° = 2\left(\frac{1}{2}\right)\left(\frac{\sqrt{3}}{2}\right) = \frac{\sqrt{3}}{2}$$

$$\cos 14\theta = 2 \cos^2 7\theta - 1$$

$$\cos \frac{8\theta}{9} = 1 - 2 \sin^2 \frac{4\theta}{9}$$

Furthermore, formula (7) implies that $\sin A \cos A = \frac{1}{2} \sin 2A$.

48 HALF–ANGLE FORMULAS

If θ is any angle, then

$$\sin \frac{\theta}{2} = \pm \sqrt{\frac{1 - \cos \theta}{2}} \tag{9}$$

$$\cos \frac{\theta}{2} = \pm \sqrt{\frac{1 + \cos \theta}{2}} \tag{10}$$

The choice of the sign in front of the radical is determined by the quadrant in which $\theta/2$ lies.

PROOF   Formula (8b) says

$$\cos 2A = 1 - 2 \sin^2 A$$

Let $A = \theta/2$; then $2A = \theta$, and

$$\cos \theta = 1 - 2 \sin^2 \frac{\theta}{2}$$

Transposing, we obtain

$$2 \sin^2 \frac{\theta}{2} = 1 - \cos \theta$$

$$\sin^2 \frac{\theta}{2} = \frac{1 - \cos \theta}{2}$$

$$\sin \frac{\theta}{2} = \pm \sqrt{\frac{1 - \cos \theta}{2}}$$

By a similar method, formula (10) can be derived from formula (8c).

Formulas (9) and (10) are called *half-angle formulas* because the angle on the left side is half the angle on the right side. To illustrate,

$$\sin 30° = \sqrt{\frac{1 - \cos 60°}{2}} = \sqrt{\frac{1 - \frac{1}{2}}{2}} = \sqrt{\frac{\frac{1}{2}}{2}} = \sqrt{\frac{1}{4}} = \frac{1}{2}$$

$$\cos 170° = -\sqrt{\frac{1 + \cos 340°}{2}}$$

(Here the minus sign is chosen because 170° is in Q II and has a negative cosine.)

$$\sin C = \pm \sqrt{\frac{1 - \cos 2C}{2}}$$

Notice that in the double-angle formulas the large angle is on the left side, while in the half-angle formulas the small angle is on the left side. In fact, the half-angle formulas are merely the double-angle formulas used backward (reading from right to left).

It is desirable to read the left side of formula (9): "the sine of half of θ" rather than to say "the sine of θ over 2" which might be construed as $\frac{\sin \theta}{2}$.

EXAMPLE 1 Prove the identity $\cos 3\theta = 4 \cos^3 \theta - 3 \cos \theta$

PROOF

$$
\begin{aligned}
\cos 3\theta & \qquad\qquad 4 \cos^3 \theta - 3 \cos \theta \\
&= \cos (2\theta + \theta) \\
&= \cos 2\theta \cos \theta - \sin 2\theta \sin \theta \\
&= (2 \cos^2 \theta - 1)* \cos \theta - (2 \sin \theta \cos \theta) \cdot \sin \theta \\
&= 2 \cos^3 \theta - \cos \theta - 2 \sin^2 \theta \cos \theta \\
&= 2 \cos^3 \theta - \cos \theta - 2(1 - \cos^2 \theta) \cos \theta \\
&= 2 \cos^3 \theta - \cos \theta - 2 \cos \theta + 2 \cos^3 \theta \\
&= 4 \cos^3 \theta - 3 \cos \theta
\end{aligned}
$$

*Formula (8c) is used because the right side involves only cos θ.

This identity is frequently used in proving that it is impossible to trisect a general angle with ruler and compass.

EXAMPLE 2 Express $\cos 20\theta$ in terms of $\sin 5\theta$.

SOLUTION Since 20θ is the double of 10θ, and 10θ is the double of 5θ, we shall employ double-angle formulas:

$$
\begin{aligned}
\cos 20\theta \\
&= 1 - 2\sin^2 10\theta \\
&= 1 - 2(2\sin 5\theta \cos 5\theta)^2 \\
&= 1 - 8\sin^2 5\theta \cos^2 5\theta \\
&= 1 - 8\sin^2 5\theta(1 - \sin^2 5\theta) \\
&= 1 - 8\sin^2 5\theta + 8\sin^4 5\theta
\end{aligned}
$$

EXAMPLE 3 By using half-angle formulas, reduce $\sin^4 A$ to an expression involving no even exponents.

SOLUTION Upon squaring both sides of formula (9), we obtain

$$
\sin^2 \frac{\theta}{2} = \frac{1 - \cos\theta}{2} = \frac{1}{2}(1 - \cos\theta)
$$

Replacing $\theta/2$ with A gives $\sin^2 A = \frac{1}{2}(1 - \cos 2A)$. This equation enables us to change the exponent from 2 to 1, with a doubling of the angle. Hence

$$
\begin{aligned}
\sin^4 A = (\sin^2 A)^2 &= [\tfrac{1}{2}(1 - \cos 2A)]^2 \\
&= \tfrac{1}{4}(1 - 2\cos 2A + \cos^2 2A)
\end{aligned}
$$

If $\cos^2 2A$ is replaced by $\frac{1}{2}(1 + \cos 4A)$, we get

$$
\sin^4 A = \tfrac{3}{8} - \tfrac{1}{2}\cos 2A + \tfrac{1}{8}\cos 4A
$$

Such transformations are needed in the integration of even powers of sines and cosines in calculus.

EXERCISE 24

1. Use a calculator or tables to show that

$$2 \sin 11° \neq \sin 22°$$
$$2 \cos 25° \neq \cos 50°$$

2. Use double-angle formulas to compute $\sin 90°$ and $\cos 90°$ from the functions of $45°$.

3. Compute $\sin 67\frac{1}{2}°$ and $\cos 67\frac{1}{2}°$ from the functions of $135°$.

4. Compute $\sin 112\frac{1}{2}°$ and $\cos 112\frac{1}{2}°$ from the functions of $225°$.

Simplify by reducing to a single term involving only one function of an angle.

5. $\sqrt{\dfrac{1 - \cos 18\theta}{2}}$

6. $-\sqrt{\dfrac{1 + \cos 320°}{2}}$

7. $8 - 16 \sin^2 B$

8. $10 \cos^2 11° - 10 \sin^2 11°$

9. $\sin \dfrac{\theta}{6} \cos \dfrac{\theta}{6}$

10. $1 - 2 \cos^2 A$

11. $1 - \sin^2 5C$

12. $20 \sin 4A \cos 4A$

13. Given $\cos A = \frac{3}{7}$ with $270° < A < 360°$, find the exact values of $\cos \dfrac{A}{2}$ and $\cos 2A$.

14. Given $\sin B = -\frac{12}{13}$ with $180° < B < 270°$, find the exact values of $\sin \dfrac{B}{2}$ and $\sin 2B$.

15. Given $\sin 10A = \frac{3}{5}$ with $0° < 10A < 90°$, find the exact values of $\sin 5A$ and $\sin 20A$.

16. Given $\cos \dfrac{C}{2} = -\dfrac{5}{8}$ with $90° < \dfrac{C}{2} < 180°$, find the exact values of $\cos C$ and $\cos \dfrac{C}{4}$.

17. Express $\sin 128°$ in terms of functions of $64°$.

18. Express $\cos 246°$ in terms of $\sin 123°$.

19. Express $\cos 130°$ in terms of a function of $260°$.

20. Express $\sin 55°$ in terms of a function of $110°$.

21. Express $\sin 3A$ in terms of a function of $6A$.

22. Express $\cos 7\theta$ in terms of a function of 14θ.

23. Express $\cos 6C$ in terms of $\cos 3C$.

24. Express $\sin 8B$ in terms of $\sin 4B$.

Identify as true or false and give reasons. Do not use a calculator or tables.

25. $1 - 4 \sin^2 \theta + 4 \sin^4 \theta = \cos^2 2\theta$

26. $\left(\dfrac{1 - \cos 160°}{2} \right)^3 = \sin^6 80°$

27. $2 \sin \dfrac{B}{2} \cos \dfrac{B}{2} = \sin B$

28. $2 \sin^2 3\theta - 1 = -\cos 6\theta$

29. $\sqrt{\dfrac{1 + \cos 200°}{2}} = \cos 80°$

30. $\cos^2 6A = \cos 12A + \sin^2 6A$

31. $1 - 2 \cos^2 8\theta = \cos (-16\theta)$

32. $\sqrt{2(1 + \cos 2B)} = \pm 2 \cos B$

33. $\cos^2 \dfrac{\theta}{5} - \sin^2 \dfrac{\theta}{5} = \dfrac{1}{5} \cos 2\theta$

34. $\sin 6A = 2 \sin 12A \cos 12A$

35. $\cos^2 \dfrac{\theta}{6} - \sin^2 \dfrac{\theta}{6} = \cos \dfrac{\theta}{3}$

36. $\sin^2 11B - \frac{1}{2} \cos 22B = \frac{1}{2}$

37. Reduce $\cos^4 \theta$ to $\frac{3}{8} + \frac{1}{2} \cos 2\theta + \frac{1}{8} \cos 4\theta$.

38. Reduce $\sin^4 2\theta$ to $\frac{3}{8} - \frac{1}{2} \cos 4\theta + \frac{1}{8} \cos 8\theta$.

39. Reduce $\sin^2 \theta \cos^2 \theta$ to $\frac{1}{8} - \frac{1}{8} \cos 4\theta$.

Prove each of the following identities by transforming one side to the other.

40. $\dfrac{1 - \tan^2 A}{1 + \tan^2 A} = \cos 2A$

41. $\csc 2\theta = \frac{1}{2} \tan \theta + \frac{1}{2} \cot \theta$

42. $\dfrac{\sin 3B}{\sin B} - \dfrac{\cos 3B}{\cos B} = 2$

43. $\dfrac{2(1 + \sin 2A)}{2 \cos 2A + \sin 4A} = \sec 2A$

44. $\dfrac{\sin^3 \theta + \cos^3 \theta}{\sin \theta + \cos \theta} = 1 - \dfrac{1}{2} \sin 2\theta$

45. $\dfrac{\cos \theta}{\cos \theta - \sin \theta} - \dfrac{\sin \theta}{\cos \theta + \sin \theta} = \sec 2\theta$

46. $\cos^2 2A + 4 \sin^2 A \cos^2 A = 1$

47. $2 \cos^2 \dfrac{A}{2} - \cos A = 1$

48. $\sin 4\theta = 8 \sin \theta \cos^3 \theta - 4 \sin \theta \cos \theta$

49. $\sin 3\theta = 3 \sin \theta - 4 \sin^3 \theta$

50. $\dfrac{2 \sin 2B}{(\cos 2B - 1)^2} = \csc^2 B \cot B$

51. $\cos 10A + \cos 5A = (2 \cos 5A - 1)(\cos 5A + 1)$

52. $\dfrac{\sin 4A}{2 \sin 2A} = \cos^4 A - \sin^4 A$

53. $\csc^2 \dfrac{\theta}{2} = \dfrac{2 \sec \theta}{\sec \theta - 1}$

54. $\dfrac{\sin 2A + \cos 2A}{\sin A \cos A} = 2 + \cot A - \tan A$

55. $\cos^2 2\theta - \cos^2 \theta = \sin^2 \theta - \sin^2 2\theta$

56. $(2 \sin^2 2A - 1)^2 = 1 - \sin^2 4A$

57. $\tan 2A = \dfrac{2 \tan A}{1 - \tan^2 A}$

58. $\tan \dfrac{\theta}{2} = \dfrac{1 - \cos \theta}{\sin \theta} = \dfrac{\sin \theta}{1 + \cos \theta}$

59. $\dfrac{1 + \sin^2 C}{1 + \cos^2 C} = \dfrac{3 - \cos 2C}{3 + \cos 2C}$

60. $\dfrac{1}{2} \sin^2 2\theta + \cos 2\theta + 2 \sin^4 \theta = 1$

61. $\dfrac{\cos A - \cos 2A}{\sin^2 A} = \dfrac{1 + 2 \cos A}{1 + \cos A}$

62. $2 \sin^2 4B(1 + \cos 8B) = \sin^2 8B$

63. $\dfrac{3 + 7 \sin 2\theta + 3 \cos 2\theta}{7 + 3 \sin 2\theta - 7 \cos 2\theta} = \cot \theta$

64. $\sin 2\theta - \cos 2\theta = \dfrac{(1 + \tan \theta)^2 - 2}{1 + \tan^2 \theta}$

65. $\dfrac{\cos \dfrac{\theta}{2} + \sin \dfrac{\theta}{2}}{\cos \dfrac{\theta}{2} - \sin \dfrac{\theta}{2}} = \sec \theta + \tan \theta$

66. $\dfrac{\sin 2B + 2 \cos 2B + 1}{\cos 2B} = \dfrac{3 \cos B - \sin B}{\cos B - \sin B}$

67. In right triangle ABC, where $C = 90°$, prove that

$$\sin \frac{A}{2} = \sqrt{\frac{c - b}{2c}}$$

68. Verify the identity in Problem 52 for $A = 30°, 45°, 60°$.
69. Verify the identity in Problem 41 for $\theta = 30°, 45°, 135°$.
70. Verify the identity in Problem 42 for $B = 30°, 45°, 60°$.
71. Verify the identity in Problem 59 for $C = 0°, 30°, 90°$.

49 PRODUCT TO SUM FORMULAS; SUM TO PRODUCT FORMULAS

It is sometimes necessary to convert a product of two trigonometric functions into a sum of two functions, and vice versa. For this reason we develop the following formulas. They are not nearly so important as the preceding 10 formulas.

When (1) and (4) are added, we get

$$\sin (A + B) + \sin (A - B) = 2 \sin A \cos B$$

or $\qquad \sin A \cos B = \tfrac{1}{2}[\sin (A + B) + \sin (A - B)]$ $\qquad$ (11)

Upon subtracting (1) and (4), we obtain

$$\cos A \sin B = \tfrac{1}{2}[\sin (A + B) - \sin (A - B)] \qquad (12)$$

Similarly, by adding and subtracting (2) and (5), we get

$$\cos A \cos B = \tfrac{1}{2}[\cos (A + B) + \cos (A - B)] \qquad (13)$$

$$\sin A \sin B = \tfrac{1}{2}[\cos (A - B) - \cos (A + B)] \qquad (14)$$

Formulas (11), (12), (13), and (14) are used to convert a product of sines and cosines into a sum or difference* of sines and cosines. They are used in certain problems in integral calculus.

When formula (11) is used backward (from right to left), it converts a sum into a product. Thus

$$\sin (A + B) + \sin (A - B) = 2 \sin A \cos B \qquad (7\text{-}5)$$

For convenience we shall change notation by making the substitutions

$$A + B = C \quad \text{and} \quad A - B = D$$

Adding these two equations and dividing by 2, we then obtain $A = \frac{1}{2}(C + D)$. Subtracting and dividing by 2, we get $B = \frac{1}{2}(C - D)$. Substituting in (7-5), we obtain

$$\sin C + \sin D = 2 \sin \tfrac{1}{2}(C + D) \cos \tfrac{1}{2}(C - D) \qquad (15)$$

Similarly,

$$\sin C - \sin D = 2 \cos \tfrac{1}{2}(C + D) \sin \tfrac{1}{2}(C - D) \qquad (16)$$

$$\cos C + \cos D = 2 \cos \tfrac{1}{2}(C + D) \cos \tfrac{1}{2}(C - D) \qquad (17)$$

$$\cos C - \cos D = -2 \sin \tfrac{1}{2}(C + D) \sin \tfrac{1}{2}(C - D) \qquad (18)$$

Formulas (15), (16), (17), and (18) are used to convert sums and differences into products. Formula (16) is usually employed in the derivation of the formula for the derivative of the sine in differential calculus.

Formula (1) should not be confused with formula (15). The former deals with the sine of the sum of two angles; the latter deals with the sum of the sines of two angles.

EXAMPLE 1 Reduce $\cos 5\theta \cos 3\theta$ to a sum.

SOLUTION Using formula (13), we get

$$\cos 5\theta \cos 3\theta = \tfrac{1}{2}[\cos (5\theta + 3\theta) + \cos (5\theta - 3\theta)]$$
$$= \tfrac{1}{2} \cos 8\theta + \tfrac{1}{2} \cos 2\theta$$

---

*The expression "sum or difference" is called the "algebraic sum." The expression $(a - b)$ may be written as $[a + (-b)]$, which is the algebraic sum of a and $-b$.

EXAMPLE 2 Prove the identity $\dfrac{\sin 7\theta - \sin 3\theta}{\cos 7\theta + \cos 3\theta} = \tan 2\theta$

PROOF $\dfrac{\sin 7\theta - \sin 3\theta}{\cos 7\theta + \cos 3\theta}$ $\Bigg|$ $\tan 2\theta$

$$= \frac{2 \cos \tfrac{1}{2}(7\theta + 3\theta) \sin \tfrac{1}{2}(7\theta - 3\theta)}{2 \cos \tfrac{1}{2}(7\theta + 3\theta) \cos \tfrac{1}{2}(7\theta - 3\theta)}$$

$$= \frac{\sin 2\theta}{\cos 2\theta}$$

$$= \tan 2\theta$$

EXERCISE 25

Express each of the following as an algebraic sum of sines and cosines:

1. $\cos 7\theta \cos 3\theta$

2. $12 \sin 9A \cos 8A$

3. $2 \sin \dfrac{3\theta}{2} \sin \dfrac{\theta}{2}$

4. $10 \cos 77° \cos 33°$

5. $8 \sin 42° \cos 107°$

6. $\sin 206° \sin 31°$

Express each of the following as a product:

7. $\cos 200° + \cos 100°$

8. $\sin 70° - \sin 10°$

9. $\cos 15A - \cos 9A$

10. $\sin B + \sin (B + 60°)$

Prove the following identities. Do not use a calculator or tables.

11. $\dfrac{\sin 250° - \sin 110°}{\cos 100° - \cos 40°} = 2$

12. $\dfrac{\sin 9\theta + \sin 7\theta}{\cos 4\theta + \cos 2\theta} = \dfrac{\sin 8\theta}{\cos 3\theta}$

13. $\dfrac{\sin 55° + \sin 15°}{\sin 155° - \sin 85°} = -2 \cos 20°$

14. $\dfrac{\cos 8\theta - \cos 2\theta}{\cos 8\theta + \cos 2\theta} = -\tan 5\theta \tan 3\theta$

15. $\cos 130° + \sin 80° = \cos 70°$

16. $\sin 18\theta \cos 14\theta - \sin 10\theta \cos 6\theta = \cos 24\theta \sin 8\theta$

17. $\sin 11A \sin 3A - \cos 7A \cos A = -\cos 10A \cos 4A$

18. $\dfrac{\cos 10\theta - \cos 6\theta}{\sin 5\theta - \sin \theta} + \dfrac{\sin 15\theta + \sin \theta}{\cos 10\theta + \cos 4\theta} = 0$

19. $\sin B + \sin 2B + \sin 3B = (1 + 2 \cos B) \sin 2B$

20. $\dfrac{\sin (A + B + C) - \sin (A - B - C)}{\cos (A + B + C) - \cos (A - B - C)} = -\cot A$

21. If $A + B + C = 180°$, prove

$$\cos A + \cos B + \cos C = 1 + 4 \sin \frac{A}{2} \sin \frac{B}{2} \sin \frac{C}{2}$$

Hint: Show that $\cos C = -\cos (A + B) = 1 - 2 \cos^2 \frac{1}{2}(A + B)$.

22. Verify formula (11) for $B = A$.

23. Verify formula (15) for $D = 0$.

/, 3, 5, 13, 17,

8 TRIGONOMETRIC EQUATIONS

50 TRIGONOMETRIC EQUATIONS

A conditional equation is an equation that does *not* hold true for all permissible values of the letters involved. (The student should review Section 26 before proceeding.) If a conditional equation involves trigonometric functions, it is called a trigonometric equation, in contrast to a trigonometric identity. *A solution of a trigonometric equation is a value of the angle that satisfies the equation.* For example, $\theta = 90°$ is a solution of the equation $\sin \theta = 1 + \cos \theta$. Any angle coterminal with $90°$ is also a solution. Thus $\theta = 450°, 810°$, etc., are also solutions. In this book, unless stated to the contrary, *to solve a trigonometric equation shall mean to find all positive (or zero) solutions less than* $360°$, that is, all θ on the interval $0° \leq \theta < 360°$, or $0 \leq \theta < 2\pi$.

51 SOLVING A TRIGONOMETRIC EQUATION

The process of finding the solutions of a trigonometric equation involves algebraic as well as trigonometric methods. There is no general rule, but the following suggestions will take care of most cases.

(*A*) *If only one function of a single angle is involved, solve algebraically for the values of the function.* Then determine the corresponding angles.

EXAMPLE 1 Solve for θ: $\quad 4 \cos^2 \theta = 3$

SOLUTION $\qquad\qquad\qquad \cos^2 \theta = \frac{3}{4}$

130

Hence

$$\cos \theta = \frac{\sqrt{3}}{2} \qquad \bigg| \qquad \cos \theta = -\frac{\sqrt{3}}{2}$$

$$\theta = 30°, 330° \qquad \bigg| \qquad \theta = 150°, 210°$$

Therefore (arranging the solutions in order of size)

$$\theta = 30°, 150°, 210°, 330°$$

EXAMPLE 2 Solve for θ: $4 \sin^2 \theta = 3 \sin \theta$

SOLUTION   Transpose $3 \sin \theta$ to make the right side 0; then factor:

$$\sin \theta (4 \sin \theta - 3) = 0$$

$$\begin{array}{l|l} \sin \theta = 0 & 4 \sin \theta - 3 = 0 \\ \theta = 0°, 180° & \sin \theta = \frac{3}{4} = 0.75 \\ & \theta = 48.59°, 131.41° \end{array}$$

Hence

$$\theta = 0°, 48.59°, 131.41°, 180°$$

The student is warned to guard against dividing both sides of this equation by the variable factor $\sin \theta$. Had this been done, the solutions $0°$ and $180°$ would have been lost.

EXAMPLE 3 Solve for θ: $\sin^2 \theta - 5 \sin \theta - 3 = 0$

SOLUTION   The left side is not factorable. The roots of the quadratic equation $ax^2 + bx + c = 0$ are

$$x = \frac{-b \pm \sqrt{b^2 - 4ac}}{2a}$$

In this case $a = 1$, $b = -5$, $c = -3$, and $x = \sin \theta$. The formula gives us

$$\sin \theta = \frac{5 \pm \sqrt{25 + 12}}{2} = \frac{5 \pm 6.0828}{2}$$

$\sin \theta = 5.5414$	$\sin \theta = -0.5414$
Impossible	(Related angle is $32.78°$)
	θ is in Q III, IV
	$\theta = 212.78°, 327.22°$

Hence

$$\theta = 212.78°, 327.22°$$

(B) *If one side of the equation is zero and the other side is factorable, set each such factor equal to zero* and solve the resulting equations.

EXAMPLE 4 Solve for θ: $\cos 2\theta \csc \theta - 2 \cos 2\theta = 0$

SOLUTION Factor the left side:

$$\cos 2\theta (\csc \theta - 2) = 0$$

$\cos 2\theta = 0$	$\csc \theta - 2 = 0$
$2\theta = 90°, 270°, 450°, 630°$	$\csc \theta = 2$
$\theta = 45°, 135°, 225°, 315°$	$\sin \theta = \frac{1}{2}$
	$\theta = 30°, 150°$

Hence

$$\theta = 30°, 45°, 135°, 150°, 225°, 315°$$

In order to solve $\cos 2\theta = 0$ for all values of θ on the interval $0° \leq \theta < 360°$, it is necessary to find all values of 2θ on the interval $0° \leq 2\theta < 720°$.

(C) *If several functions of a single angle are involved,* use the fundamental relations to *express everything in terms of a single function.* Then proceed as in (A).

EXAMPLE 5 Solve for θ: $\sin^2 \theta - \cos^2 \theta - \cos \theta = 1$

SOLUTION Replace $\sin^2 \theta$ with $1 - \cos^2 \theta$; collect terms:

$$2 \cos^2 \theta + \cos \theta = 0$$
$$\cos \theta \, (2 \cos \theta + 1) = 0$$

$\cos \theta = 0$	$2 \cos \theta + 1 = 0$
$\theta = 90°, 270°$	$\cos \theta = -\frac{1}{2}$
	(Related angle is $60°$)
	θ is in Q II, III
	$\theta = 120°, 240°$

Hence

$$\theta = 90°, 120°, 240°, 270°$$

EXAMPLE 6   Solve for θ: $\sec \theta = \tan \theta - 1$

SOLUTION   Replace $\sec \theta$ with $\pm \sqrt{1 + \tan^2 \theta}$; square both sides:

$$1 + \tan^2 \theta = \tan^2 \theta - 2 \tan \theta + 1$$
$$2 \tan \theta = 0 \qquad \tan \theta = 0$$

Hence

$$\theta = 0°, 180°$$

Inasmuch as we squared the equation, we may have introduced some extraneous roots. Consequently, we must check these values in the original equation.

Check for $\theta = 0°$:

$$\sec 0° = \tan 0° - 1$$
$$1 = -1 \qquad \text{False}$$

Check for $\theta = 180°$:

$$\sec 180° = \tan 180° - 1$$
$$-1 = -1 \qquad \text{True}$$

Therefore $\theta = 0°$ is an *extraneous* root and $\theta = 180°$ is a true root. The only solution of the equation is $\theta = 180°$.

(D) If several angles are involved, use the fundamental identities to *express everything in terms of a single angle.* Then proceed as in (*C*).

EXAMPLE 7 Solve for θ: $\cos 2\theta = 3 \sin \theta + 2$

SOLUTION This equation involves two angles. It is not convenient to replace $\sin \theta$ with a function of 2θ because this would introduce the radical

$$\pm \sqrt{\frac{1 - \cos 2\theta}{2}}$$

It is better to replace $\cos 2\theta$ with one of its three forms. Since the right side involves only $\sin \theta$, we choose the form $\cos 2\theta = 1 - 2 \sin^2 \theta$ in order to reduce everything immediately to the same function of a single angle. This gives us

$$1 - 2 \sin^2 \theta = 3 \sin \theta + 2$$

or $$2 \sin^2 \theta + 3 \sin \theta + 1 = 0$$

Factor:

$$(2 \sin \theta + 1)(\sin \theta + 1) = 0$$

$$\begin{array}{c|c} \sin \theta = -\frac{1}{2} & \sin \theta = -1 \\ \theta = 210°, 330° & \theta = 270° \end{array}$$

Hence*

$$\theta = 210°, 270°, 330°$$

EXERCISE 26

Solve the following equations for θ on the interval $0° \leq \theta < 360°$.

1. $2 \sin^2 \theta = 1$ **2.** $\csc^2 \theta = 4$

---

*All the solutions of this equation could be written as $210° + k \cdot 360°$, $270° + k \cdot 360°$, $330° + k \cdot 360°$, where k is an integer. Another form is $270° + k \cdot 360°$, $\left[k + \frac{(-1)^{k+1}}{6} \right] \cdot 180°$, where k is an integer.

3. $4 \cos^2 \theta = 1$

4. $\tan^2 \theta = 3$

5. $\sin (\theta - 7°) = -\dfrac{\sqrt{3}}{2}$

6. $\cos (\theta + 10°) = \dfrac{\sqrt{2}}{2}$

7. $(\sqrt{2} \sin \theta - 1)(2 \cos \theta + \sqrt{3}) = 0$

8. $(\sqrt{3} \sec \theta - 2)(2 \sin \theta + \sqrt{3}) = 0$

9. $(\csc \theta - 2)(2 \cos \theta - \sqrt{3}) = 0$

10. $(2 \cos \theta - 1)(\sqrt{3} \csc \theta + 2) = 0$

11. $\tan^2 \theta = \tan \theta$

12. $\csc^3 \theta = 2 \csc \theta$

Solve the following equations for θ on the interval $0 \le \theta < 2\pi$.

13. $2 \cos^2 \theta - \cos \theta = 1$

14. $2 \cos \theta \sin \theta + 2 \cos \theta + \sin \theta + 1 = 0$

15. $2 \sin^2 \theta + 3 \sin \theta + 1 = 0$

16. $2 \cos^2 \theta - 9 \cos \theta - 5 = 0$

17. $\cos \theta = \sqrt{3} \sin \theta$

18. $2 \sin \theta \tan \theta - \sqrt{3} \tan \theta = 0$

19. $6 \sec^2 \theta + 5 \sec \theta + 1 = 0$

20. $2 \sin^2 \theta + 7 \cos \theta - 5 = 0$

21. $\cos \theta + \sec \theta = 1.9$

22. $3 \sec \theta \csc \theta = \csc \theta$

23. $\cos 2\theta = 1 - \sin \theta$

24. $\sin 4\theta - \sqrt{3} \cos 2\theta = 0$

25. $\sin 2\theta + \sqrt{2} \sin \theta = 0$

26. $\cos 4\theta = 7 \cos 2\theta + 8$

27. $\sin \theta = \sin \dfrac{\theta}{2}$

28. $\cos 2\theta - 2 \cos^2 \theta = 0$

Solve the following equations for θ on the interval $0° \le \theta < 360°$.

29. $\cos 6\theta + \sin^2 3\theta = 0$

30. $\sin 2\theta + 2 \cos^2 \dfrac{\theta}{2} = 1$

31. $\sin^2 \theta = 6 \cos \theta + 6$

32. $2 \cos^2 \theta + 7 \sin \theta = 5$

33. $\tan \theta + \cot \theta = 2 \csc \theta$

34. $\sin^2 \theta = 3 \cos^2 \theta$

35. $\dfrac{1}{\cot \theta + \tan \theta} = \sin \theta \cos \theta$

36. $\dfrac{\tan 5\theta - \tan 2\theta}{1 + \tan 5\theta \tan 2\theta} = \dfrac{1}{\sqrt{3}}$

37. $\tan \theta + 1 = -\sqrt{2} \sec \theta$

38. $\sin \theta - 2 = \sqrt{3} \cos \theta$

39. $\csc \theta = \cot \theta - 1$

40. $\cos \theta - \sin \theta = \sqrt{2}$

Solve for θ on the interval $0° \le \theta < 360°$. In some instances a calculator or Table 1 will be needed. Find θ to the nearest tenth of a degree.

41. $30 \cos^2 \theta - 11 \cos \theta + 1 = 0$

42. $\tan^2 \theta + \tan \theta - 6 = 0$

43. $9 \sin^4 \theta - 19 \sin^2 \theta + 2 = 0$

44. $6 \cos \theta + 1 = \sec \theta$

45. $27 \sec \theta \tan \theta - 8 \csc \theta \cot \theta = 0$

46. $\cos^2 \theta = 6 \sin \theta + 5$

47. $\sec^2 \theta - 5 \tan \theta + 3 = 0$

48. $\cot^2 \theta - 7 \csc \theta + 7 = 0$

49. $4 \sin \theta + 3 \cos \theta = 2.5$

Hint: Express the left side in the form $k \sin (\theta + H)$.

50. $-12 \sin \theta + 5 \cos \theta = 6.5$

51. $24 \sin \theta + 7 \cos \theta = 20$

52. $\sqrt{3} \sin \theta - \cos \theta = 1$

53. $\cos 3\theta = \sin (\theta + 10°)$

Hint: Transpose one term to the other side. Replace $\sin (\theta + 10°)$ with $\cos (80° - \theta)$. Express the difference as a product. As a partial check, sketch on the same system of coordinates the graphs of $y_1 = \cos 3\theta$ and $y_2 = \sin (\theta + 10°)$; then approximate the value of θ at each point of intersection.

54. $\sin 4\theta - \sin (2\theta - 30°) = 0$

55. $\sin 7\theta + \sin \theta = 6 \cos 3\theta$

56. $\cos 3\theta + \cos \theta = \cos 2\theta$

57. $4 \tan \theta = \sqrt{3} \sec^2 \theta$

58. $3 \sec 2\theta + 2 \sin^2 \theta + 1 = 0$

59. $\sqrt{2}\sin\dfrac{\theta}{2} + \sqrt{1-\cos\theta} = \sqrt{2}$

60. $\tan(45° - \theta) = 1 - \sin 2\theta$

61. $\cot\theta + \sin^2\theta - \cos^2\theta\cot\theta = 0$

62. $3\sqrt{2}\cos\dfrac{\theta}{2} + \sqrt{1+\cos\theta} = -\sqrt{6}$

9 GRAPHICAL METHODS

52 THE GRAPH OF $y = a \sin bx$

The graph of $y = \sin x$ (Figure 49) is a wave that starts at 0, goes up to 1, drops back to 0, then to -1, and finally climbs back up to 0. (We are discussing the changes that take place in y as x varies from 0 to 2π.) The amplitude of the wave (i.e., the maximum distance from the x axis) is 1. The period, or wavelength, which is the same as the period* of the function $\sin x$, is 2π. The graph of $y = 3 \sin x$ (Figure 60) is another wave, with amplitude 3 and period 2π. For any value of x, the magnitude of y in $y = 3 \sin x$ is 3 times the value of y in $y = \sin x$. The coefficient 3 merely *stretches* the *height* of the sine curve by the multiple 3 without affecting its period.

The graph of $y = \sin 2x$ is another sine wave† (Figure 61) with amplitude 1 but with period π. As x varies from 0 to $\pi/4$, $2x$ changes from 0 to $\pi/2$, and $\sin 2x$ increases from 0 to 1. The coefficient 2 *compresses* the curve *horizontally* by the multiple $\frac{1}{2}$. The period of the function $\sin 2x$ is half the period of $\sin x$. (Things happen twice as fast.)

To generalize and combine our two previous observations, we may say that

> The graph of $y = a \sin bx$ is a sine wave with amplitude a and period $\dfrac{2\pi}{b}$.

---

*See Section 37.

† We reserve the term "sine *curve*" for the graph of $y = \sin x$.

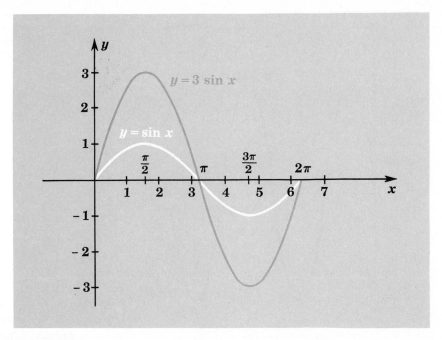

FIGURE 60

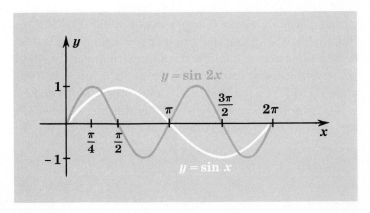

FIGURE 61

A similar statement can be made for $y = a \cos bx$. Figure 62 shows the graph of $y = 3 \sin 2x$.

53 THE GRAPH OF $y = a \sin (bx + c)$

Consider the equation $y = \sin (x + 30°)$. When $x = 0°, y = \sin 30° = \frac{1}{2}$. If $x = 60°, y = \sin 90° = 1$. As x varies from $0°$ to $360°$, y takes on the

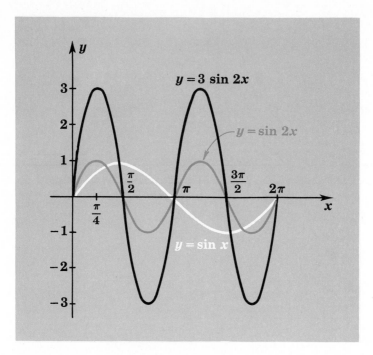

FIGURE 62

same values assumed by y in the equation $y = \sin x$ *in the same order but starting at a different place.* The graph of $y = \sin(x + 30°)$ may be obtained (Figure 63) by shifting the graph of $y = \sin x$ to the *left* 30°, or $\pi/6$ units. If the reader does not follow the argument, he or she should plot as many points as necessary using $y = \sin(x + 30°)$. Suggested values to assign to x are 0°, 60°, 150°, 240°, 330°, and others if needed.

The graph of $y = \sin(x - \pi/2)$ is the sine curve shifted $\pi/2$ units to the *right* (Figure 63). Another approach is to use the formula for

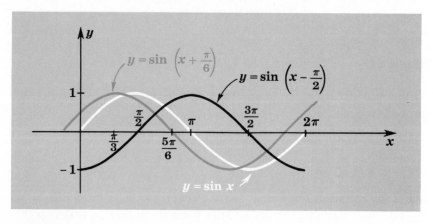

FIGURE 63

$\sin (A - B)$*and find $\sin (x - \pi/2) = -\cos x$. The graph of $y = -\cos x$ is merely the reflection of the graph of $y = \cos x$ in the x axis. (Place a two-sided mirror on the x axis. The graph of $y = -\cos x$ is the image of the graph of $y = \cos x$.) For any given value of x, the y of $y = -\cos x$ is the negative of the y of $y = \cos x$.

The graph of $y = \sin (x + k)$ may be obtained by shifting the graph of $y = \sin x$ a distance $|k|$ to the $\left\{ \begin{array}{l} left\ if\ k\ is\ positive \\ right\ if\ k\ is\ negative \end{array} \right\}$. *If you are in doubt, check with one point; set $x = 0$. The quantity $|k|$ is called the phase displacement.*

Let us now consider the equation $y = 2.5 \sin (2x + \pi/3)$, which is equivalent† to $y = 2.5 \sin 2(x + \pi/6)$. First, sketch the graph of $y = \sin x$ (Figure 64). Second, *compress* the curve *horizontally* by the multiple $\frac{1}{2}$ to get

---

*Formula (4), Section 45.

† Two equations are equivalent if every solution of one equation is a solution of the other, and vice versa.

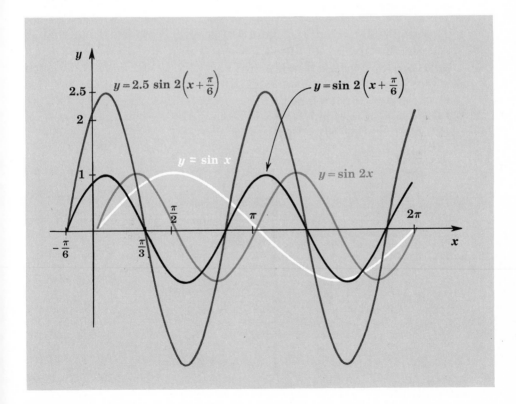

FIGURE 64

the graph of $y = \sin 2x$ (Figures 61 and 64). Third, *shift* this curve $\pi/6$ units to the *left* to obtain the graph of $y = \sin 2(x + \pi/6)$. Finally, *stretch* this curve vertically to 2.5 times its original height to get the graph of $y = 2.5 \sin 2(x + \pi/6)$ (Figure 64). Again the reader is encouraged to check the result by assigning a few values to x in the original equation, computing the corresponding values of y, and then plotting the points. Some of the strategic points are $(0, 2.2)$, $(\pi/12, 2.5)$,* $(\pi/3, 0)$, $(7\pi/12, -2.5)$,* $(5\pi/6, 0)$. For the function $2.5 \sin (2x + \pi/3)$, the amplitude is 2.5, the period is $2\pi/2 = \pi$, and the phase displacement is $\pi/3 \div 2 = \pi/6$.

The function $a \sin (bx + c)$ has amplitude a, period $2\pi/b$, and phase displacement $|c/b|$. The graph of $y = a \sin (bx + c)$ may be obtained from the graph of $y = a \sin bx$ by shifting it horizontally $|c/b|$ units: to the *left* if c/b is *positive*, to the *right* if c/b is *negative*. A similar statement holds true for $a \cos (bx + c)$.

54 THE GRAPH OF $y = \sin^n x$

The graph of $y = \sin^3 x$ (Figure 65) may be obtained from that for $y = \sin x$ by "cubing all the y's." If $x = 0$, $\sin x = 0$ and $\sin^3 x = 0$, but for $x = \pi/6$, $\sin x = \frac{1}{2}$ whereas $\sin^3 x = \frac{1}{8}$.

---

*The maximum value of y is achieved when $\sin (2x + \pi/3) = 1$; $2x + \pi/3 = \pi/2$; $x = \pi/12$. The minimum value of y occurs when $2x + \pi/3 = 3\pi/2$; $x = 7\pi/12$.

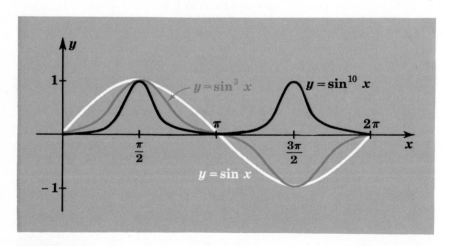

FIGURE 65

Using a similar procedure, we find that the graph of $y = \sin^{10} x$ (Figure 65) does not fall below the x axis because 10 is even and therefore $y \geq 0$ for all values of x.

Consideration of the graphs of $y = \sin^n x$, where n is a positive integer greater than 1, more properly belongs in differential calculus, where it is demonstrated that each of these curves passes *horizontally* through the points $(0, 0)$, $(\pi/2, 1)$, $(\pi, 0)$, etc. Moreover, in integral calculus we learn that the area under one arch of the sine curve (i.e., the first-quadrant area beneath the curve $y = \sin x$ from $x = 0$ to $x = \pi$) is 2 square units. This seems reasonable because the circumscribing rectangle has altitude 1 and base π; its area is π square units. Furthermore, from calculus, the area under one arch of $y = \sin^3 x$ is $1\frac{1}{3}$ square units, and the area under one arch of $y = \sin^{10} x$ is $63\pi/256$ square units.

55 SKETCHING CURVES BY COMPOSITION OF y COORDINATES

If the graphs of $y = f(x)$ and $y = g(x)$ are drawn to the same scale on the same set of axes, the graph of

$$y = f(x) + g(x)$$

can be sketched by the process of *composition of y coordinates*. For any value of x, we can determine the y of the equation $y = f(x) + g(x)$ by finding graphically the *algebraic* sum of the y's of the two equations $y = f(x)$ and $y = g(x)$. After a suitable number of points have been located by "adding the heights of the given curves," we connect them with a smooth curve to get the required graph.

EXAMPLE 1 Graph the equation $y = x + \sin x$.

SOLUTION First draw the graphs of $y = x$ and $y = \sin x$ on the same axes (x being measured in radians). Place a straightedge parallel to the y axis at M_1. Use compasses to add the segments $M_1 S_1$ and $M_1 R_1$. The sum is $M_1 P_1$, thus locating P_1. To get point P_2, add the negative segment $M_2 S_2$ to the positive segment $M_2 R_2$; their *algebraic* sum is $M_2 P_2$ (see Figure 66).

EXAMPLE 2 Graph the equation $y = \sin x + \sin 2x$.

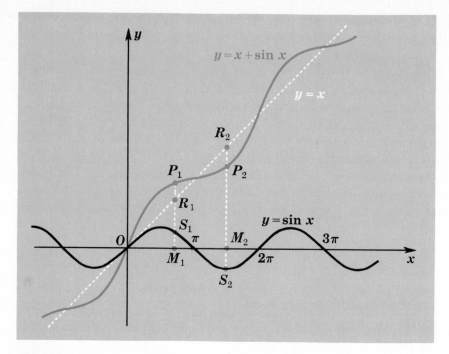

FIGURE 66

SOLUTION After graphing the equations $y = \sin x$ and $y = \sin 2x$, we use the process of composition of y coordinates (Figure 67).

56 THE GRAPH OF $y = a \sin x + b \cos x$

One plan of attack in drawing the graph of $y = a \sin x + b \cos x$ is first to sketch (on the same coordinate system) the graphs of $y = a \sin x$ and $y = b \cos x$ and then to use the method of composition of y coordinates. A much shorter way, however, is to reduce $y = a \sin x + b \cos x$ to the form $y = k \sin (x + H)$, the graph of which is a sine wave with amplitude k, period 2π, and phase displacement $|H|$.

EXAMPLE Sketch the graph of $y = 4 \sin x + 3 \cos x$.

SOLUTION From the example of Section 46, we have $4 \sin x + 3 \cos x = 5 \sin (x + 0.64)$. Hence the graph of $y = 4 \sin x + 3 \cos x$ is the same as the graph (Figure 68) of $y = 5 \sin (x + 0.64)$.

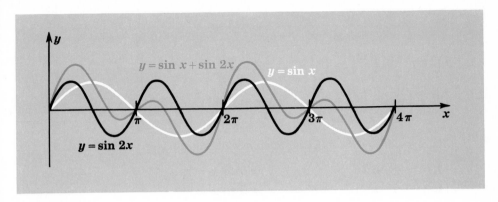

FIGURE 67

EXERCISE 27

Find the period and amplitude of the following functions of x:

1. $\sin 5x$ **2.** $3 \sin 4x$ **3.** $\frac{1}{2} \sin \pi x$ **4.** $\sin \dfrac{\pi x}{5}$

5. $7 \cos \dfrac{x}{4}$ **6.** $\tan \dfrac{5\pi x}{6}$ **7.** $9 \tan \dfrac{4x}{5}$ **8.** $6 \cos 8x$

Find the period and amplitude of the function defined by the first equation in each of the following problems. Give the position of its graph (number of units right or left) relative to the graph of the second equation.

9. $y = 3.5 \sin (4x + 4)$ $y = 3.5 \sin 4x$

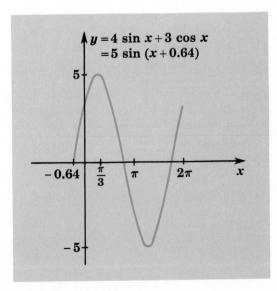

FIGURE 68

10. $y = 8 \sin \left(\dfrac{3x}{2} - \dfrac{15\pi}{4} \right)$ $y = 8 \sin \dfrac{3x}{2}$

11. $y = \sin \left(\dfrac{\pi x}{3} - 2\pi \right)$ $y = \sin \dfrac{\pi x}{3}$

12. $y = 5 \sin \left(\dfrac{x}{2} + \dfrac{3}{2} \right)$ $y = 5 \sin \dfrac{x}{2}$

Sketch graphs of the following equations on the specified intervals.

13. $y = 4 \sin 3x$ $x = 0$ to $x = \pi$

14. $y = 5 \cos 2x$ $x = 0$ to $x = \pi$

15. $y = \cos \dfrac{\pi x}{2}$ $x = 0$ to $x = 4$

16. $y = 3 \sin \dfrac{x}{3}$ $x = 0$ to $x = 6\pi$

17. $y = 6 \cos (x - \pi)$ $x = 0$ to $x = 2\pi$

18. $y = \sin \left(x - \dfrac{\pi}{3} \right)$ $x = 0$ to $x = 2\pi$

19. $y = 2 \sin \left(x + \dfrac{\pi}{6} \right)$ $x = 0$ to $x = 2\pi$

20. $y = 1.5 \sin \left(x + \dfrac{\pi}{4} \right)$ $x = 0$ to $x = 2\pi$

21. $y = 5 \sin \left(\dfrac{x}{2} + 30° \right)$ $x = 0°$ to $x = 720°$

22. $y = 3 \sin \left(\dfrac{x}{3} + 30° \right)$ $x = 0°$ to $x = 1080°$

23. $y = 4 \sin (3x - 30°)$ $x = 0°$ to $x = 180°$

24. $y = \sin (2x - 60°)$ $x = 0°$ to $x = 360°$

25. $y = \tan \left(x + \dfrac{\pi}{2} \right)$ $x = 0$ to $x = 2\pi$

26. $y = \sin^5 x$ $x = 0$ to $x = 2\pi$

27. $y = \cos^4 x$ $x = 0$ to $x = 2\pi$

28. $y = \tan 2x$ $x = 0$ to $x = \pi$

Use composition of y coordinates in sketching graphs of the following equations.

29. $y = -1 + \sin x$ 30. $y = -x + \cos x$

31. $y = \dfrac{x}{\pi} + \sin x$ 32. $y = x^2 + \cos x$

33. $y = \cos x + \cos 3x$

34. $y = \sin x + \frac{1}{2}\sin 2x$

35. $y = \frac{1}{2}\cos 2x - \frac{1}{2}$

Graph the following equations after changing them to the form $y = k \sin(x + H)$.

36. $y = 3 \sin x + 4 \cos x$

37. $y = \sin x - \cos x$

38. $y = \sqrt{3} \sin x + \cos x$

39. $y = 2.4 \sin x + 0.7 \cos x$

40. (*a*) Graph $y = \cos^2 x$.

    (*b*) Graph $y = \frac{1}{2} + \frac{1}{2}\cos 2x$.

    (*c*) Why are the graphs identical?

10 LOGARITHMS

57 THE USES OF LOGARITHMS

Long, drawn-out computations have always been sheer drudgery for people who deal with numbers. Some relief was obtained in the middle 1600s after a Scotsman, John Napier, and an Englishman, Henry Briggs, invented logarithms. By using logarithms they were able to replace the operations of multiplication and division with addition and subtraction. As a result, logarithms have been widely used in trigonometric computations for the past three hundred years. (The slide rule was a consequence of the invention of logarithms.) In recent years the use of logarithms for computation has been replaced in large part by the electronic calculator. Today the analytic aspects of logarithms and their applications in higher mathematics and science are far more important than their use in performing numerical computations. For these reasons, the theory and properties of logarithms will be discussed in this chapter, but their computational capabilities will be treated in the Appendix.

58 SOME LAWS OF EXPONENTS

A logarithm, as we shall see later, is an exponent. Accordingly we shall first review the following laws of exponents.

1. $a^m \cdot a^n = a^{m+n}$ *Example:* $2^3 \cdot 2^4 = 2^7$

2. $(a^m)^n = a^{mn}$ *Example:* $(2^3)^4 = 2^{12}$

3. $\dfrac{a^m}{a^n} = a^{m-n}$ *Example:* $\dfrac{2^8}{2^2} = 2^6$

4. $a^{m/n} = (\sqrt[n]{a})^m = \sqrt[n]{a^m}$ *Example:* $8^{4/3} = (\sqrt[3]{8})^4 = 2^4 = 16$

5. $a^0 = 1$ *Example:* $(\frac{2}{3})^0 = 1$

6. $a^{-n} = \dfrac{1}{a^n}$ *Example:* $3^{-2} = \dfrac{1}{3^2} = \dfrac{1}{9}$

Although these laws are true for all values of m and n and for positive and negative values of a, we shall have occasion to use them only for positive values of a.

59 DEFINITION OF LOGARITHM

The logarithm of a number to a given base is the exponent which must be placed on the base to produce the number.*

Thus the logarithm of 9 to the base 3 (written $\log_3 9$) is 2 because $3^2 = 9$.

ILLUSTRATIONS

$\log_2 8 = 3$	because	$2^3 = 8$
$\log_7 1 = 0$	because	$7^0 = 1$
$\log_3 \frac{1}{3} = -1$	because	$3^{-1} = \frac{1}{3}$
$\log_2 \dfrac{1}{16} = -4$	because	$2^{-4} = \dfrac{1}{2^4} = \dfrac{1}{16}$
$\log_{25} 5 = \frac{1}{2}$	because	$(25)^{1/2} = \sqrt{25} = 5$
$\log_8 4 = \frac{2}{3}$	because	$8^{2/3} = (\sqrt[3]{8})^2 = 2^2 = 4$
$\log_6 6 = 1$	because	$6^1 = 6$

The definition of logarithm implies that

if $\log_b N = x$

then $b^x = N$

These two equations, the former logarithmic and the latter exponential, are equivalent. They say the same thing in two different ways. We shall assume in further discussions that N is a positive number and that b is a

---

*This exponent need not be rational. For example, $\log_{10} 2$ is an irrational number which, rounded off to five decimal places, becomes 0.30103. If the exponent $\dfrac{30,103}{100,000}$ is placed on 10, the result is 2 *approximately*.

positive number different from 1. (Explain the necessity for such restrictions.) But x may be any real number: positive, negative, or zero.

Since, by the definition of logarithm, $\log_b N$ is the exponent which must be placed on b to produce N, it follows that if $\log_b N$ is applied as an exponent to the base b, then the result must be N:

$$b^{\log_b N} = N$$

EXERCISE 28

Find the value of each of the following logarithms:

1. $\log_4 1$
2. $\log_{16} 2$
3. $\log_5 125$
4. $\log_{1/7} \frac{1}{49}$
5. $\log_2 64$
6. $\log_{23} 23$
7. $\log_{32} \frac{1}{8}$
8. $\log_4 \frac{1}{4}$
9. $\log_{27} 9$
10. $\log_{1/8} 16$
11. $\log_{13} 1$
12. $\log_9 27$
13. $\log_{1/5} 25$
14. $\log_4 256$
15. $\log_{81} 27$
16. $\log_{32} \frac{1}{2}$

Find the unknown, N, b, or x, in the following:

17. $\log_b 6 = -1$
18. $\log_4 N = 0$
19. $\log_{15} N = 1$
20. $5^{\log_5 12} = x$
21. $e^{\log_e 9} = x$
22. $\log_b \frac{1}{27} = -3$
23. $\log_{1/2} 16 = x$
24. $\log_b 243 = -5$
25. $\log_{256} N = -\frac{3}{4}$
26. $\log_7 7^{31} = x$
27. $\log_b 4^7 = 7$
28. $\log_{64} N = \frac{5}{6}$

Express as a logarithmic equation.

29. $9^{5/2} = 243$
30. $81^{-1/2} = \frac{1}{9}$
31. $64^{-1/6} = \frac{1}{2}$
32. $m^p = q$

Express as an exponential equation.

33. $\log_b a = c$
34. $\log_{1024} 16 = \frac{2}{5}$
35. $\log_{121} 11 = \frac{1}{2}$
36. $\log_{7.6} 1 = 0$

Identify as true or false and give reasons.

37. $\log_{1000} \dfrac{\sqrt{10}}{10} = -\dfrac{1}{6}$

38. $\log_{27} \sqrt[4]{3} = \frac{1}{12}$

39. $\log_5 25^r = 2r$

40. $\log_{32} 2^t = \dfrac{t}{5}$

60 PROPERTIES OF LOGARITHMS

As a consequence of the definition of logarithm, we have three properties or laws of logarithms. They are used in computations involving logarithms.

PROPERTY 1   The logarithm of a product is equal to the sum of the logarithms of the factors; i.e.,

$$\log MN = \log M + \log N^*$$

PROOF   Let $x = \log_a M$ and $y = \log_a N$

Express in exponential form: $M = a^x$ and $N = a^y$

Multiply the equations: $MN = a^x a^y = a^{x+y}$

Change to logarithmic form: $\log_a MN = x + y$
$\log_a MN = \log_a M + \log_a N$

The proof is similar for a product of more than two factors.

ILLUSTRATIONS $\log 35 = \log 5 \cdot 7 = \log 5 + \log 7$
$\log 30 = \log 2 \cdot 3 \cdot 5 = \log 2 + \log 3 + \log 5$

PROPERTY 2   The logarithm of a fraction is equal to the logarithm of the numerator minus the logarithm of the denominator; i.e.,

$$\log \frac{M}{N} = \log M - \log N$$

---

*The base is the same for all the logarithms.

PROOF   Let $x = \log_a M$ and $y = \log_a N$

Express in exponential form: $M = a^x$ and $N = a^y$

Divide the equations: $\dfrac{M}{N} = \dfrac{a^x}{a^y} = a^{x-y}$

Change to logarithmic form: $\log_a \dfrac{M}{N} = x - y$

$$\log_a \dfrac{M}{N} = \log_a M - \log_a N$$

ILLUSTRATIONS $\log \frac{2}{3} = \log 2 - \log 3$

$$\log \frac{6}{35} = \log \frac{2 \cdot 3}{5 \cdot 7} = \log 2 + \log 3 - (\log 5 + \log 7)$$

PROPERTY 3 The logarithm of the kth power of a number is equal to k times the logarithm of the number; i.e.,

$$\log N^k = k \log N$$

PROOF   Let $x = \log_a N$

Express in exponential form: $N = a^x$

Raise to the kth power: $N^k = (a^x)^k = a^{kx}$

Change to logarithmic form: $\log_a N^k = kx$
$\log_a N^k = k \log_a N$

ILLUSTRATIONS $\log 8 = \log 2^3 = 3 \log 2$

$\log \sqrt{3} = \log 3^{1/2} = \frac{1}{2} \log 3$

$\log \dfrac{125}{49} = \log \dfrac{5^3}{7^2} = \log 5^3 - \log 7^2 = 3 \log 5 - 2 \log 7$

Note   Since $\sqrt[r]{M} = M^{1/r}$, $\log \sqrt[r]{M} = \dfrac{1}{r} \log M$.

EXERCISE 29

Given $\log_{10} 2 = 0.30$, $\log_{10} 3 = 0.48$, and $\log_{10} 7 = 0.85$, find the following logarithms to two-decimal-place accuracy. (Remember $\log_{10} 10 = 1$.)

1. $\log_{10} 14$
2. $\log_{10} \frac{2}{7}$
3. $\log_{10} \frac{3}{2}$
4. $\log_{10} 60$
5. $\log_{10} \sqrt{3}$
6. $\log_{10} 27$
7. $\log_{10} 32$
8. $\log_{10} \sqrt[4]{7}$
9. $\log_{10} \dfrac{2^{10}}{7}$
10. $\log_{10} \sqrt[3]{21}$
11. $\log_{10} 7\sqrt{3}$
12. $\log_{10} \left(\frac{7}{2}\right)^8$

Express as a single logarithm. (Assume all logarithms have the same base.)

13. $\log (a + b) - \log b$
14. $\log P + n \log (1 + i)$
15. $6 \log a + \frac{1}{2} \log (b^2 + c^2)$
16. $\frac{1}{5} \log a - \frac{2}{3} \log b^6$
17. $4 \log x + \frac{1}{3} \log y + \frac{1}{6} \log z$
18. $\log r + \frac{1}{2} (\log s - \log t)$
19. $\log (x^2 - y^2) - \log (x - y) + \log z$
20. $\log r + 2 \log s - \log t - 3 \log w$

Identify as true or false and give reasons. (In each equation the base is assumed to be the same for all the logarithms.)

21. $(\log a)^2 = 2 \log a$
22. $\log \dfrac{1}{x} = -\log x$
23. $\dfrac{x}{y} = \log x - \log y$
24. $\sqrt[3]{a} = \frac{1}{3} \log a$
25. $\log N = \dfrac{a}{b} \log N^{b/a}$
26. $\log \sqrt{ab} = \sqrt{\log a + \log b}$
27. $3 \log (\log a) = \log (\log a)^3$
28. $9^{\log_9 a + 5 \log_9 b} = ab^5$
29. $7^{(\log_7 x)/2} = \sqrt{x}$
30. $10^{2 + \log_{10} 5} = 500$
31. $6^{4 \log_6 a - \log_6 b} = \dfrac{a^4}{b}$
32. $\dfrac{1}{3} \log r + \dfrac{1}{4} \log s^2 - \dfrac{1}{6} \log \sqrt{t} = \dfrac{1}{12} \log \dfrac{r^4 s^6}{t}$

33. $\log \dfrac{xy^2}{z} = \dfrac{\log x + 2 \log y}{\log z}$

34. $\log (\sec \theta + \tan \theta) = -\log (\sec \theta - \tan \theta) \qquad 0 < \theta < \dfrac{\pi}{2}$

61 SYSTEMS OF LOGARITHMS

There are only two important systems of logarithms in use today. The *natural*, or *napierian*, system employs the base e, where e is approximately 2.71828. This system is encountered in calculus and higher mathematics. The *common*, or *Briggs*, system employs the base 10. This system is most convenient for computation because our number system uses the base 10.

62 THE GRAPHS OF $y = a^x$ AND $y = \log_a x$

If $a > 1$, the graph of $y = a^x$ is basically the same as that of $y = 2^x$ (Figure 69), which was obtained by use of the following table.

x	$\ldots$	-3	-2	-1	0	1	2	3	$\ldots$
$y = 2^x$	$\ldots$	$\frac{1}{8}$	$\frac{1}{4}$	$\frac{1}{2}$	1	2	4	8	$\ldots$

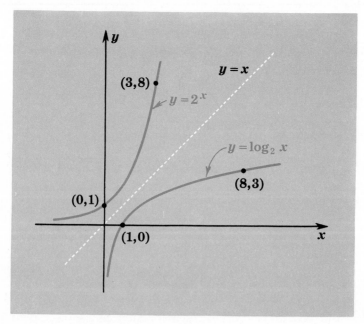

FIGURE 69

If $a > 1$, the graph of $y = \log_a x$ is essentially the same as that of $y = \log_2 x$ (Figure 69), which was drawn by use of the following table.

x	$\cdots$	$\frac{1}{8}$	$\frac{1}{4}$	$\frac{1}{2}$	1	2	4	8	$\cdots$
$y = \log_2 x$	$\cdots$	-3	-2	-1	0	1	2	3	$\cdots$

Notice that the function 2^x increases as x increases. This function could be called "the doubling function" because whenever x increases by 1, the function doubles its size. ($2^{x+1} = 2 \cdot 2^x$.)

Observe that the function $\log_2 x$ also increases as x increases. Notice that $\log_2 x$ is not defined for $x < 0$. A negative number does not have a (real) logarithm.

The equation $y = \log_2 x$ is equivalent to the equation $x = 2^y$, which is obtainable from $y = 2^x$ by merely interchanging x and y. If (a, b) is a point on the graph of $y = 2^x$, then (b, a) is a point on the graph of $y = \log_2 x$. It can be shown that if a mirror were placed on the line $y = x$, the image of the point (a, b) would be the point (b, a). Hence the graph of $y = \log_2 x$ is the reflection of the graph of $y = 2^x$ in the line $y = x$. As we shall learn in Chapter 12, the functions 2^x and $\log_2 x$ are *inverse functions*.

63 LOGARITHMIC EQUATIONS*

A logarithmic equation is an equation that contains the logarithm of some expression involving the unknown. Such an equation can usually be solved by rewriting it in one of the following two forms:

1. $\log M = \log N$ which implies $M = N$

or

2. $\log_b N = w$ which implies $N = b^w$

EXAMPLE 1 Solve for x: $\log x + 3 \log a = \frac{1}{2} \log b$

SOLUTION   We shall rewrite the equation in the form $\log M = \log N$ and then state that $M = N$. (If two numbers have equal logarithms, the numbers must be equal.)

---

*Exponential equations are discussed in Section 91 in the Appendix.

$$\log x + 3 \log a = \tfrac{1}{2} \log b$$
$$\log x + \log a^3 = \log b^{1/2}$$
$$\log a^3 x = \log \sqrt{b}$$
$$a^3 x = \sqrt{b}$$

$$x = \frac{\sqrt{b}}{a^3}$$

COMMENT   A common mistake is to say that if $\log A + \log B = \log C$, then $A + B = C$. This is incorrect because $\log A + \log B \neq \log (A + B)$. The proper way to handle this situation is to replace $\log A + \log B$ with $\log AB$. Then $\log AB = \log C$; $AB = C$.

EXAMPLE 2   Solve for x: $\log_{10} x + \log_{10} a = 3 + 4 \log_{10} b$

SOLUTION 1   We shall rewrite the equation in the form $\log_{10} N = w$ and then assert that $N = 10^w$ (definition of logarithm). Then

$$\log_{10} x + \log_{10} a - \log_{10} b^4 = 3$$

$$\log_{10} \frac{xa}{b^4} = 3$$

$$\frac{xa}{b^4} = 10^3$$

$$x = \frac{1000b^4}{a}$$

This result may be checked by equating the logarithms of the two sides of the last equation and then showing that the resulting equation is equivalent to the given equation.

SOLUTION 2   Replace 3 with $\log_{10} 1000$ and proceed as in Example 1. Then

$$\log_{10} x + \log_{10} a = \log_{10} 1000 + \log_{10} b^4$$
$$\log_{10} xa = \log_{10} 1000b^4$$
$$xa = 1000b^4$$

$$x = \frac{1000b^4}{a}$$

EXERCISE 30

Solve the following equations for x:

1. $\log (1 - x) + \log (6 - x) = \log 14$
2. $\log x + \log (x - 2) = \log (8 - 4x)$
3. $\log (x - 1) + \log (x + 1) = \log (6x - 8)$
4. $2 \log (x + 6) = \log (x + 9) + \log 4$
5. $\log_{10} (9x + 14) = 1 + \log_{10} (2x - 3)$
6. $\log_2 (21x + 25) = 5 + \log_2 (x - 3)$
7. $\log_3 (x + 7) + \log_3 (x - 17) = 4$
8. $\log_{10} (7x + 5) + 2 \log_{10} 2 = 2 + \log_{10} 3$
9. $y = \log_b x + \log_b (x - 8)$
10. $y = \dfrac{1}{2} \log_b \dfrac{a + cx}{f + gx}$

Sketch graphs of the following equations:

11. $y = 3^x$
12. $y = \log_x 3$
13. $y = (\tfrac{1}{2})^x$
14. $y = \log_{1/2} x$

11 OBLIQUE TRIANGLES

64 INTRODUCTION

A triangle that does not contain a right angle is called an oblique triangle. Since the ratio of two sides of an *oblique* triangle does not represent a function of an angle of the triangle, additional formulas are needed for solving oblique triangles. These formulas are the law of sines, the law of cosines, the law of tangents, and the half-angle formulas.

The six parts of a triangle are its three sides a, b, c, and the opposite angles A, B, C, respectively. If three parts, at least one of which is a side, are given, the remaining parts can be determined. For convenience, we shall divide the possibilities into the following four cases:

SAA: Given one side and two angles.

SSA: Given two sides and the angle opposite one of them.

SAS: Given two sides and the included angle.

SSS: Given three sides.

65 THE LAW OF SINES

In any triangle the sides are proportional to the sines of the opposite angles:

$$\frac{a}{\sin A} = \frac{b}{\sin B} = \frac{c}{\sin C}$$

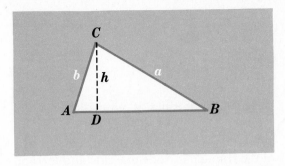

FIGURE 70

PROOF   Consider the two cases: all angles acute (Figure 70) and one angle obtuse (Figure 71). Let h be the perpendicular from C to AB (or AB extended).

In either case

$$\sin B = \frac{h}{a} \qquad \text{hence} \qquad h = a \sin B$$

Also* $\sin A = \dfrac{h}{b}$ hence $h = b \sin A$

Equate the two values of h: $a \sin B = b \sin A$

Divide by $\sin A \sin B$: $\dfrac{a}{\sin A} = \dfrac{b}{\sin B}$

Similarly, $\dfrac{a}{\sin A} = \dfrac{c}{\sin C}$

---

*In Figure 71, $\sin A = \sin \angle CAD = \dfrac{h}{b}$.

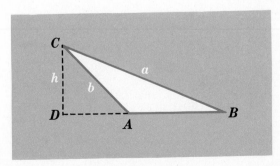

FIGURE 71

Therefore

$$\frac{a}{\sin A} = \frac{b}{\sin B} = \frac{c}{\sin C}$$

The law of sines is equivalent to the following three equations: $\frac{a}{\sin A} = \frac{b}{\sin B}$, $\frac{a}{\sin A} = \frac{c}{\sin C}$, $\frac{b}{\sin B} = \frac{c}{\sin C}$. If the three given parts of the triangle include one side and the opposite angle, then one of these equations will involve three known quantities and one unknown. We can solve for this unknown by using ordinary algebraic methods. Thus, if a, c, and C are the known parts, we can find A by solving the equation

$$\frac{\sin A}{a} = \frac{\sin C}{c} \qquad \sin A = \frac{a \sin C}{c}$$

The law of sines is well adapted to the use of logarithms because it involves only ratios and products. *Whenever four parts of a triangle are known* (the fourth part may have been obtained from the other three), *we can use the law of sines to find the remaining two parts.* Why?

66 APPLICATIONS OF THE LAW OF SINES: *SAA*

When one side and two angles of a triangle are known, we can immediately find the third angle from the relation $A + B + C = 180°$. The remaining two sides can then be found by using the law of sines.

EXAMPLE 1 Solve the triangle ABC, given $a = 20$, $A = 30°$, $B = 40°$.

SOLUTION   (See Figure 72.)

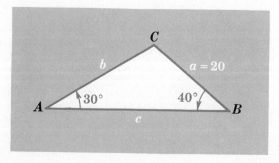

FIGURE 72

(1)
$$C = 180° - (30° + 40°)$$
$$= 110°$$

(2) To find b, use

$$b = \frac{a \sin B}{\sin A} = \frac{20 \sin 40°}{\sin 30°} = \frac{20(0.6428)}{0.5000} \doteq 26$$

(3) To find c, use

$$c = \frac{a \sin C}{\sin A} = \frac{20 \sin 110°}{\sin 30°} = \frac{20(0.9397)}{0.5000} \doteq 38$$

The values of b and c are rounded off to two-figure accuracy. The results can be checked by the law of cosines (Section 68) or by finding c with the formula $c = \dfrac{b \sin C}{\sin B}$.

EXAMPLE 2 Solve the triangle ABC, given $b = 191$, $A = 55.7°$, $C = 81.5°$ (see Figure 73).

SOLUTION PLAN

(1)
$$B = 180° - (55.7° + 81.5°)$$
$$= 42.8°$$

(2)
$$a = \frac{b \sin A}{\sin B} = \frac{191 \sin 55.7°}{\sin 42.8°}$$

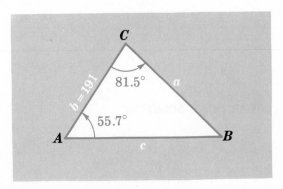

FIGURE 73

$$(3) \qquad c = \frac{b \sin C}{\sin B} = \frac{191 \sin 81.5°}{\sin 42.8°}$$

CALCULATOR SOLUTION

$$(2) \qquad a = \frac{191 \sin 55.7°}{\sin 42.8°}$$

Using an algebraic calculator with the switch set for degrees, we press

$$\boxed{191} \;\; \boxed{\times} \;\; \boxed{55.7} \;\; \boxed{\sin} \;\; \boxed{=} \;\; \boxed{\div} \;\; \boxed{42.8} \;\; \boxed{\sin} \;\; \boxed{=}$$

The displayed result is 232.2272333. Rounding off to three-figure accuracy, we obtain

$$a = 232$$

Using an RPN calculator, in degree mode, we press

$$\boxed{191} \;\; \boxed{\text{ENTER}} \;\; \boxed{55.7} \;\; \boxed{\sin} \;\; \boxed{\times} \;\; \boxed{42.8} \;\; \boxed{\sin} \;\; \boxed{\div}$$

In this case the displayed result is 232.2272332, which differs slightly from the algebraic calculator answer. (Observe that two different calculators may produce results that are not quite identical.) Once again, however, the answer to three significant figures is

$$a = 232$$

(3) Using the same procedure, we find $c = 278.0255317$ or 278.0255318, each of which, after rounding off, becomes

$$c = 278$$

LOGARITHMIC SOLUTION

$$(2) \qquad a = \frac{191 \sin 55.7°}{\sin 42.8°}$$

An outline of the logarithmic solution is

$$
\begin{array}{rl}
\log 191 = & \\
\log \sin 55.7° = & \\
\hline
\log \text{ numerator} = & \quad \text{Add} \\
\log \sin 42.8° = & \\
\hline
\log a = & \quad \text{Subtract} \\
a = &
\end{array}
$$

The student should fill in the outline and find $a = 232$.

$$
\textbf{(3)} \qquad c = \frac{191 \sin 81.5°}{\sin 42.8°}
$$

After preparing the outline and then filling in the numbers, we have

$$
\begin{array}{rl}
\log 191 = & 2.2810 \\
\log \sin 81.5° = & 9.9952 - 10 \\
\hline
\log \text{ num.} = & 12.2762 - 10 \\
\log \sin 42.8° = & 9.8322 - 10 \\
\hline
\log c = & 2.4440 \\
c = & 278
\end{array}
\qquad
\begin{array}{l}
\text{A} \\
\\
\text{S}
\end{array}
$$

The results may be checked by finding c again using $c = \dfrac{a \sin C}{\sin A}$ or by using the law of tangents (Section 95).

EXERCISE 31

Solve the following triangles by using a calculator or tables:

1. $c = 600$, $A = 42.7°$, $C = 68.5°$
2. $a = 2.0$, $B = 108°$, $C = 39°$
3. $b = 53.7$, $A = 74.1°$, $C = 26.3°$
4. $c = 614$, $A = 51.6°$, $B = 95.2°$
5. $a = 7.89$, $B = 23.0°$, $C = 114.6°$

6. $b = 0.160$, $A = 87.2°$, $B = 49.7°$

7. $c = 952$, $B = 93.5°$, $C = 38.6°$

8. $a = 4.75$, $A = 79.3°$, $C = 56.4°$

9. $b = 0.003416$, $B = 65.80°$, $C = 84.00°$

10. $c = 6932$, $A = 12.84°$, $B = 67.16°$

11. $a = 8.269$, $A = 30.55°$, $B = 71.62°$

12. $b = 3784$, $A = 82.71°$, $C = 54.36°$

13. An observer at A looks due east and sees a UFO with an angle of elevation of 40°. At the same instant, another observer, 1.0 mi due west of A, looks due east and sights the same UFO with an angle of elevation of 25°. Find the distance between A and the UFO. How far is the UFO above the ground?

14. Vienna is 360 km N 89° E from Munich. The bearing of Milan from Munich is S 32° W. The bearing of Vienna from Milan is N 61° E. How far is Milan from Munich? From Vienna?*

15. A motorboat heads N 79.0° W on a river that flows due south. The boat travels S 86.0° W with a speed of 400 meters/min. Find the speed of the current and the speed of the boat in still water.

16. The largest angle of a triangle is twice the size of the smallest angle. Is the largest side twice as large as the smallest side? Find c if $A = 40.5°$, $C = 81.0°$, $a = 100$.

17. Show that if $C = 90°$, the law of sines gives the definition of the sine of an acute angle.

18. Why does the law of sines fail to solve the *SAS* and *SSS* cases?

67 THE AMBIGUOUS CASE: SSA

If two sides and the angle opposite one of them are given, the triangle is not always uniquely determined. With the given parts, we may be able to construct two triangles, or only one triangle, or no triangle at all. Because of the possibility of two triangles, this is usually called the ambiguous case.

To avoid unnecessary confusion, we shall use A to designate the given

---

*Ignore the curvature of the earth and assume only two-place accuracy. Get distances to the nearest multiple of 10 km.

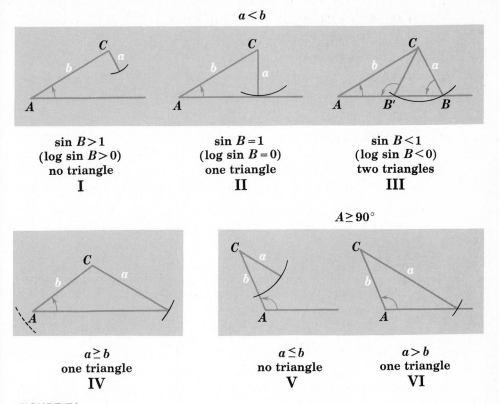

FIGURE 74

angle, a to represent the opposite side, and b to indicate the other given side. Construct angle A and lay off b as an adjacent side, thus fixing the vertex C. With C as center and a as radius, strike an arc cutting the other side adjacent to A. Figure 74 illustrates the various possibilities.

The last three cases (diagrams IV, V, and VI) can be quickly identified by merely noting the relative sizes of a and b. If A is acute and $a < b$, it is necessary to begin the solution before we can state how many triangles are possible. To do this, use the law of sines:

$$\frac{\sin B}{b} = \frac{\sin A}{a} \qquad \sin B = \frac{b \sin A}{a}$$

After determining the value of $\sin B$ (or $\log \sin B$ with logarithms), we can definitely classify our problem as one of the various types. It is frequently possible to determine the number of solutions by merely constructing a figure to scale.

ILLUSTRATION 1   Given $a = 100$, $b = 70$, $A = 80°$. This is case IV because $a > b$. There is one triangle.

ILLUSTRATION 2   Given $a = 40$, $b = 42$, $A = 110°$. This is case V because $A > 90°$ and $a < b$. There is no triangle.

ILLUSTRATION 3   Given $a = 80$, $b = 100$, $A = 54°$. Since A is acute and $a < b$, this is case I, II, or III. A carefully constructed figure leaves some doubt as to whether a is long enough to reach the horizontal side of angle A. Using the law of sines, we find

$$\sin B = \frac{b \sin A}{a} = \frac{100(0.8090)}{80} = 1.0112$$

Since no angle can have a sine greater than 1, B is impossible and there is no triangle. Had logarithms been used, we should have found $\log \sin B = 0.0049$. Since $\log \sin B > 0$, $\sin B > 1$,* B is impossible, and there is no triangle. This is case I.

EXAMPLE   Solve the triangle ABC, given $a = 48.8$, $b = 69.2$, $A = 37.2°$.

SOLUTION   A scale drawing (Figure 75) indicates that there are two triangles. Using the law of sines, we get

---

*Recall that a number greater than 1 has a logarithm greater than 0.

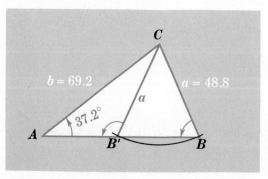

FIGURE 75

$$\frac{\sin B}{b} = \frac{\sin A}{a}$$

$$\sin B = \frac{b \sin A}{a} = \frac{69.2 \sin 37.2°}{48.8}$$

Using a calculator, we find $B = 59.0°$.
   If logarithms are used, we obtain

$$
\begin{array}{rl}
\log 69.2 = & 1.8401 \\
\log \sin 37.2° = & \underline{9.7815 - 10} \\
\log \text{num.} = & 11.6216 - 10 \\
\log 48.8 = & \underline{1.6884} \\
\log \sin B = & 9.9332 - 10 \\
B = & 59.0°
\end{array}
\quad
\begin{array}{l}
\text{A} \\[24pt]
\text{S}
\end{array}
$$

Continuing with the solution of large triangle ABC (Figure 76), we find

$$C = 180° - (37.2° + 59.0°)$$
$$= 83.8°$$

To find c, use

$$\frac{c}{\sin C} = \frac{a}{\sin A} \qquad c = \frac{a \sin C}{\sin A} = \frac{48.8 \sin 83.8°}{\sin 37.2°}$$

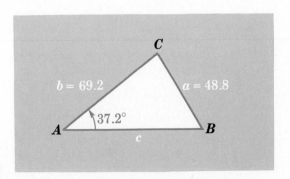

FIGURE 76

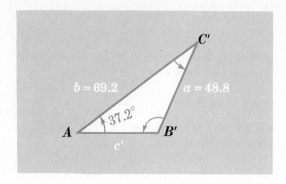

FIGURE 77

If a calculator is used, the result is $c = 80.2$. Logarithmic computation gives

$$\begin{array}{rl} \log 48.8 = & 1.6884 \\ \log \sin 83.8° = & 9.9975 - 10 \\ \hline \log \text{num.} = & 11.6859 - 10 \\ \log \sin 37.2° = & 9.7815 - 10 \\ \hline \log c = & 1.9044 \\ c = & 80.2 \end{array} \quad \begin{array}{l} \text{A} \\[10pt] \text{S} \end{array}$$

Hence the computed parts of triangle ABC are $B = 59.0°$, $C = 83.8°$, $c = 80.2$.

We shall now solve the small triangle $AB'C'$. Since triangle BCB' (Figure 75) is isosceles, $\angle BB'C = B$. Hence

$$B' = 180° - B = 180° - 59.0° = 121.0°$$

Also (see Figure 77),

$$C' = 180° - (37.2° + 121.0°) = 21.8°$$

Using the law of sines with either a calculator or logarithms, we find $c' = 30.0$.

Hence the computed parts of triangle $AB'C'$ are $B' = 121.0°$, $C' = 21.8°$, $c' = 30.0$.

EXERCISE 32

Draw a figure and solve all possible triangles.

1. $a = 20.0$, $b = 30.0$, $A = 40.0°$
2. $a = 700$, $b = 800$, $A = 64.3°$
3. $a = 0.512$, $b = 0.448$, $A = 79.1°$
4. $a = 40.9$, $b = 53.6$, $A = 42.7°$
5. $a = 670$, $b = 684$, $A = 98.6°$
6. $a = 39.7$, $b = 79.4$, $A = 30.0°$
7. $a = 1.73$, $b = 2.58$, $A = 27.4°$
8. $a = 0.0822$, $b = 0.0935$, $A = 61.8°$
9. $a = 9175$, $c = 7460$, $A = 112.35°$
10. $a = 3.890$, $c = 6.512$, $A = 18.58°$
11. $b = 1980$, $c = 1942$, $C = 96.97°$
12. $b = 920.7$, $c = 861.3$, $B = 23.24°$

13. Chapel Hill, N.C., is 330 mi from Zanesville, Ohio, and 220 mi from Charleston. The bearing of Chapel Hill from Zanesville is S 30° E. The bearing of Charleston from Zanesville is S 13° E. How far is Charleston from Zanesville? *

14. On a hillside that makes an angle of 21° with the horizontal, prevailing winds have caused a 6.0-meter tree to lean *uphill*. When the elevation of the sun is 43°, the tree's shadow is 9.0 meters long and falls straight *down* the slope. What angle does the tree make with the vertical?

15. The earth is 93 million mi from the sun. The planet Venus is 67 million mi from the sun. How far is Venus from the earth when the sun has just set in the west and Venus is 25° above the horizon in the west? (Get the result to the nearest multiple of a million miles.)

16. The pilot of a helicopter wants to fly on a course of 160°. Her plane has a cruising speed of 120 km/h in still air. In what direction should she head her plane if a 30-km/h wind is blowing in the direction 110°? Find the ground speed of the plane.

---

*Ignore the curvature of the earth and assume only two-place accuracy. Get distances to the nearest multiple of 10 mi.

68 THE LAW OF COSINES

The square of any side of a triangle is equal to the sum of the squares of the other two sides minus twice their product times the cosine of the included angle:

$$a^2 = b^2 + c^2 - 2bc \cos A \qquad \cos A = \frac{b^2 + c^2 - a^2}{2bc}$$

$$b^2 = c^2 + a^2 - 2ca \cos B \qquad \cos B = \frac{c^2 + a^2 - b^2}{2ca}$$

$$c^2 = a^2 + b^2 - 2ab \cos C \qquad \cos C = \frac{a^2 + b^2 - c^2}{2ab}$$

PROOF

(1) If all the angles are acute: Draw CD perpendicular to AB. Then (see Figure 78)

$$
\begin{aligned}
a^2 &= h^2 + \overline{DB}^2 \\
&= h^2 + (c - AD)^2 \\
&= \underbrace{h^2 + \overline{AD}^2}_{b^2} + c^2 - 2cAD \\
&= \quad b^2 \quad\; + c^2 - 2c(b \cos A) \\
a^2 &= b^2 + c^2 - 2bc \cos A
\end{aligned}
$$

(2) If one angle is obtuse: Draw CD perpendicular to AB extended. Notice that if AB is considered a positive directed segment, then AD is negative in Figure 79 while DB is negative in Figure 80. With this understanding, the proofs

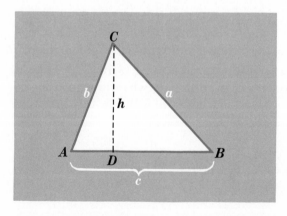

FIGURE 78

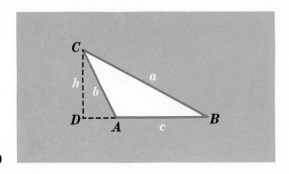

FIGURE 79

for Figures 79 and 80 are exactly the same as that for Figure 78. The student should verify this.

This is a general proof because a may be used to represent any side of the given triangle. The other two forms of the law of cosines can be obtained from the first form by the method of *cyclic permutation* in which

a is changed to b,	A is changed to B,
b is changed to c,	B is changed to C,
c is changed to a,	C is changed to A.

If $A = 90°$, $\cos A = 0$, and $a^2 = b^2 + c^2 - 2bc \cos A$ reduces to the form $a^2 = b^2 + c^2$. We conclude that the pythagorean theorem is a special case of the law of cosines; i.e., the law of cosines is a generalization of the pythagorean theorem.

69 APPLICATIONS OF THE LAW OF COSINES: *SAS* AND *SSS*

The law of cosines is used in many geometric problems, one of which is the solution of triangles. Since three sides and one angle are involved in

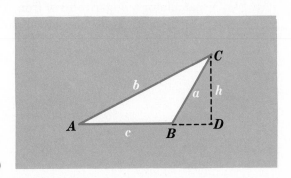

FIGURE 80

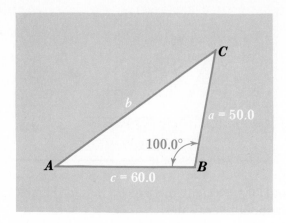

FIGURE 81

every form of the law of cosines, it can be used to solve the *SAS* and *SSS* cases described in Section 64.

EXAMPLE 1 Solve the triangle *ABC*, given $a = 50.0$, $c = 60.0$, $B = 100.0°$ (see Figure 81).

SOLUTION This is the two sides and included angle case, *SAS*. Hence

$$(1) \qquad \begin{aligned} b^2 &= c^2 + a^2 - 2ca \cos B \\ &= 60^2 + 50^2 - 2(60)(50) \cos 100° \\ &= 3600 + 2500 - 6000(-0.1736) \\ &= 7141.6 \\ b &= 84.5 \end{aligned}$$

Notice that $\cos 100° = -\cos 80° = -0.1736$.

The square root of 7141.6 may be obtained by using a calculator or logarithms or from Table 4, with interpolation.

(2) To find *A*, use the law of sines:

$$\sin A = \frac{a \sin B}{b} = \frac{50 \sin 100°}{b}$$

$$A = \text{Sin}^{-1}\left(\frac{50 \sin 100°}{b}\right)$$

$$= 35.6°$$

(3)
$$C = \text{Sin}^{-1}\left(\frac{60 \sin 100°}{b}\right)$$

$$= 44.4°$$

Check $A + B + C = 35.6° + 100.0° + 44.4° = 180.0°$

COMMENTS If you are using a calculator to find A and C, be sure to use the displayed value of b (84.50969806) rather than the rounded-off value (84.5).

This example illustrates three-figure accuracy in the given data and the computed results.

In the SAS *case*, if the given angle is acute, one of the other angles may be obtuse. If c is the longest side of a triangle and if $c^2 > a^2 + b^2$, then C must be an obtuse angle.

EXAMPLE 2 Given $a = 8.20$, $b = 5.10$, $c = 4.10$, find the largest and smallest angles (see Figure 82).

SOLUTION The largest and smallest angles lie opposite the largest and smallest sides, respectively.

(1) To find A, use

$$\cos A = \frac{b^2 + c^2 - a^2}{2bc}$$

$$= \frac{5.1^2 + 4.1^2 - 8.2^2}{2(5.1)(4.1)}$$

$$A = \text{Cos}^{-1}\left[\frac{5.1^2 + 4.1^2 - 8.2^2}{2(5.1)(4.1)}\right]$$

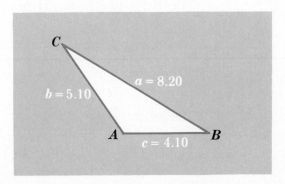

FIGURE 82

Using an algebraic logic calculator, in degree mode, we press

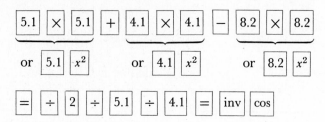

When the displayed result is rounded off to the nearest tenth of a degree, we get

$$A = 125.7°$$

(If $-1 \leq u < 0$, then the calculator is programmed to give the angle in Q II in evaluating $\text{Cos}^{-1} u$. See Section 73.)

With an RPN calculator in degree mode, press

| 5.1 | ENTER | 5.1 | × | 4.1 | ENTER | 4.1 | × | + | 8.2 |

| ENTER | 8.2 | × | − | 2 | ENTER | 5.1 | × | 4.1 | × | ÷ |

| inv | cos |

Rounding off the displayed result to three-figure accuracy, we obtain $A = 125.7°$.

If a calculator is not available, use Table 4 to square the two-digit numbers. (Since $5.1 = 51/10$, then $5.1^2 = (51)^2/100 = 26.01$, etc.) Hence

$$\cos A = \frac{26.01 + 16.81 - 67.24}{2(5.1)(4.1)} = \frac{-24.42}{41.82}$$

$$\doteq -0.5839$$

(The related angle is 54.3°.)

$$A = 180° - 54.3° = 125.7°$$

(2) To find C, use

$$\cos C = \frac{a^2 + b^2 - c^2}{2ab} = \frac{8.2^2 + 5.1^2 - 4.1^2}{2(8.2)(5.1)}$$

$$\doteq 0.91392$$

$$C = 23.9°$$

Using a calculator, we get the same answer.

EXERCISE 33

Solve the triangles in Problems 1 to 8 by using a calculator or tables. Use a calculator to solve the triangles in Problems 9 to 12.

1. $a = 5.00,\ b = 7.00,\ c = 10.00$
2. $a = 11,\ b = 12,\ c = 20$
3. $a = 49.0,\ b = 39.0,\ c = 16.0$
4. $a = 130,\ b = 480,\ c = 430$
5. $b = 80,\ c = 30,\ A = 106°$
6. $a = 1.5,\ c = 8.0,\ B = 62°$
7. $a = 510,\ b = 620,\ C = 97.0°$
8. $b = 2.00,\ c = 2.50,\ A = 115.0°$

9. $a = 5050,\ b = 7071,\ c = 8082$
10. $a = 7428,\ b = 5913,\ C = 59.66°$
11. $a = 486,\ c = 937,\ B = 25.1°$
12. $a = 43.2,\ b = 45.6,\ c = 60.6$

13. The minute hand of a campus clock is 2.0 meters long. Its hour hand has a length of 1.2 meters. Find the distance between the tips of the hands at 8 o'clock. (Consider the numbers to be exact.)
14. Two ships leave a port at 1:00 P.M. The first ship travels in the direction S 40° E with a speed of 10 mi/h. The second ship's speed is 14 mi/h. If the two ships are 23 mi apart at 2:30 P.M., find the direction in which the second ship is traveling.
15. Pomona, Calif., is 310 mi S 36° E from Modesto, Calif. Las Vegas, Nev., is 210 mi from Pomona and 340 mi from Modesto. What is the

bearing of Las Vegas from Modesto? From Pomona? (Las Vegas, of course, lies north and east from the Modesto-Pomona line.) *

16. Grand Rapids, Mich., is 380 mi N 39° E from St. Louis, Mo. Vincennes, Ind., is 150 mi N 89° E from St. Louis. Find the bearing and the distance of Grand Rapids from Vincennes. *

17. An airplane heads east (direction 90°) with an air speed of 220 km/h. A wind speed of 46 km/h causes the plane to have a ground speed of 240 km/h. Find the wind direction and the plane's course.

18. A river flows southeast (S 45° E) at 5.5 km/h. If a motorboat is to move eastward at 19 km/h, find the proper heading and the boat's cruising speed in still water (that is, its "water speed").

70 SUMMARY

We have seen that, of the four problems listed in Section 64, two (*SAA* and *SSA*) can be solved with the law of sines, while the other two (*SAS* and *SSS*) require the law of cosines. If you use a calculator that deals with trigonometric functions, you need only these two laws. But if you use logarithms in making calculations, you will find that the law of cosines is not well adapted to logarithmic computation [$\log (R + S) \neq \log R + \log S$]. For this reason it can occasionally be a matter of some importance to take advantage of alternative formulas—the "law of tangents" and the "half-angle formulas"—which are better suited to computation using logarithms and frequently produce more accurate results. These formulas, together with an outline to assist the student in choosing the best plan of attack, are presented in the Appendix, Sections 94 to 98. But if extreme accuracy is not of the essence, all four cases can be treated by using only the law of sines and the law of cosines. The student should recall that the law of sines can be used to complete the solution as soon as four parts are known.

71 THE AREA OF A TRIANGLE

I. The area of a triangle is equal to one-half the product of any two sides times the sine of the included angle:

$$K = \tfrac{1}{2}bc \sin A = \tfrac{1}{2}ca \sin B = \tfrac{1}{2}ab \sin C$$

---

*Ignore the curvature of the earth and assume only two-place accuracy. Get angles to the nearest degree. Get distances to the nearest multiple of 10 mi.

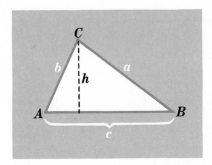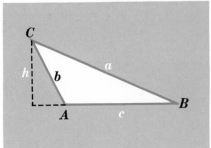

FIGURE 83

PROOF Let h be the altitude from C to AB (Figure 83).

$$\text{Area } \triangle ABC = K = \tfrac{1}{2}ch = \tfrac{1}{2}bc \sin A$$

since $h = b \sin A$. The other forms can be obtained by cyclic permutation.

II. The area of triangle ABC is

$$K = \sqrt{s(s-a)(s-b)(s-c)}$$

where $s = \tfrac{1}{2}(a + b + c)$

PROOF Since $K = \tfrac{1}{2}bc \sin A$,

$$K^2 = \frac{1}{4}b^2c^2 \sin^2 A = \frac{b^2c^2}{4}(1 - \cos^2 A)$$

$$= \frac{bc}{2}(1 + \cos A)\frac{bc}{2}(1 - \cos A)$$

Using the law of cosines, we get

$$K^2 = \frac{bc}{2}\left(1 + \frac{b^2 + c^2 - a^2}{2bc}\right)\frac{bc}{2}\left(1 - \frac{b^2 + c^2 - a^2}{2bc}\right)$$

$$= \frac{2bc + b^2 + c^2 - a^2}{4} \cdot \frac{2bc - b^2 - c^2 + a^2}{4}$$

$$= \frac{(b + c)^2 - a^2}{4} \cdot \frac{a^2 - (b - c)^2}{4}$$

$$= \frac{b + c + a}{2} \cdot \frac{b + c - a}{2} \cdot \frac{a - b + c}{2} \cdot \frac{a + b - c}{2}$$

Let $s = \dfrac{1}{2}(a + b + c)$; then $\dfrac{b + c - a}{2} = s - a$, etc.

Hence

$$K^2 = s(s - a)(s - b)(s - c)$$
$$K = \sqrt{s(s - a)(s - b)(s - c)}$$

III. The area of triangle ABC is

$$K = \frac{a^2 \sin B \sin C}{2 \sin A} = \frac{b^2 \sin C \sin A}{2 \sin B} = \frac{c^2 \sin A \sin B}{2 \sin C}$$

PROOF   We start with $K = \frac{1}{2}ab \sin C$ and use the law of sines to replace b with $\dfrac{a \sin B}{\sin A}$. This gives $K = \dfrac{a^2 \sin B \sin C}{2 \sin A}$. This formula should be used when one side and two angles are given.

EXERCISE 34

Find the areas of the following triangles:

1. $a = 35$, $b = 20$, $C = 18°$
2. $b = 8.00$, $c = 5.00$, $A = 37.3°$
3. $c = 10$, $A = 6°$, $C = 150°$
4. $a = 30$, $B = 130°$, $C = 20°$
5. $a = 41$, $b = 416$, $c = 425$ (consider the numbers as exact)
6. $a = 13$, $b = 20$, $c = 25$ (consider the numbers as exact)
7. $a = 6.04$, $c = 9.67$, $B = 53.13°$
8. $a = 32.66$, $b = 87.80$, $C = 61.65°$
9. $b = 931$, $B = 29.5°$, $C = 86.9°$
10. $c = 6.835$, $A = 74.26°$, $B = 31.44°$
11. $a = 57.20$, $b = 74.33$, $c = 66.47$
12. $a = 700$, $b = 581$, $c = 593$

13. Prove that the area of any quadrilateral is equal to one-half the product of its diagonals times the sine of the included angle.

14. The area of a triangle is 60.0 square meters. Two angles of the triangle are 39.0° and 75.0°. Find the length of the side included by these angles.

15. Use area formula I to derive an expression for the area of an equilateral triangle. Check your result by using (*a*) formula II, (*b*) formula III.

16. Use area formula II to show that the area of an isosceles triangle with sides *a*, *a*, and *b* is $\dfrac{b}{4}\sqrt{4a^2 - b^2}$.

EXERCISE 35 Miscellaneous problems

For each of the eight following problems, (a) draw the triangle, (b) write the equation that should be used in solving for the required part, and (c) find the required part.

1. $b = 16$, $A = 72°$, $C = 41°$; find *c*.
2. $a = 600$, $c = 800$, $B = 95.0°$; find *b*.
3. $b = 37$, $c = 44$, $B = 53°$; find *C*.
4. $a = 4.0$, $b = 8.0$, $c = 7.0$; find *B*.
5. $a = 910$, $b = 730$, $c = 820$; find *A*.
6. $a = 7.77$, $b = 6.00$, $B = 34.5°$; find *A*.
7. $b = 580$, $c = 250$, $A = 80.6°$; find *a*.
8. $c = 52.6$, $B = 29.0°$, $C = 67.3°$; find *b*.

9. Two straight roads intersect at an angle of 82.0°. A bus on one road is 200 meters from the intersection. A truck on the other road is 300 meters from the crossing. Find the distance between the vehicles.

10. A speedboat race is held over a triangular course, the first leg of which is 200 meters in a westerly direction. The second and third legs, of 150 and 300 meters, respectively, lie to the north of the first leg. In what direction do the boats move on the second leg?

11. From point *A*, the angle of elevation of a mountain peak is 31.5°. From point *B*, in the same horizontal plane with *A* and 475 meters closer to the base of the mountain, the angle of elevation of the peak is 40.7°. Find the height of the peak above the level of *A* and *B*.

12. The longer side of a parallelogram is 3.570 meters. The longer diagonal is 4.680 meters. The diagonals intersect at an angle of 121.00°. Find the length of the other diagonal.

13. A sailboat is moving 3.0 mi/h along the shoreline in the direction N 21° W. A motorboat with a speed of 10.0 mi/h is due east of the sailboat. In what direction should the motorboat travel if it is to overtake the sailboat in a minimum amount of time?

14. In order to find the width AB of a lake, the distance from A to a tower T is measured and found to be 404.0 meters. If angle BAT is 77.50° and angle BTA is 63.48°, how wide is the lake?

15. The diagonals of a parallelogram are 30 and 60. If one side of the parallelogram is 40, find the other side. (Consider the numbers as exact.)

16. Find the magnitude and the direction of the resultant of forces of 20 kg acting N 45° W and 60 kg acting N 15° E.

17. Houston is 230 mi S 22° E from Dallas. The bearing of Baton Rouge from Dallas is S 64° E. The bearing of Baton Rouge from Houston is N 79° E. How far is Baton Rouge from Dallas? From Houston? *

18. Fort Wayne, Ind., is 260 mi from Aliquippa, Pa., and 140 mi from Columbus. The bearing of Fort Wayne from Aliquippa is N 83° W. The bearing of Columbus from Aliquippa is S 72° W. How far is Columbus from Aliquippa? *

19. Atlanta, Ga., is 240 mi N 2° W from Tallahassee, Fla. Montgomery, Ala., is 180 mi N 42° W from Tallahassee. Find the bearing and the distance of Atlanta from Montgomery. *

20. Des Moines, Iowa, is 250 mi N 74° W from Peoria, Ill. Rolla, Mo., is 310 mi from Des Moines and 260 mi from Peoria. What is the bearing of Rolla from Peoria? From Des Moines? * (Rolla, of course, lies south of the Des Moines–Peoria line.)

21. The distance between the centers of two intersecting circles is 10.0 cm. The radii of the circles are 5.0 cm and 7.0 cm, respectively. Find the acute angle made by the tangent lines at a point of intersection.

22. An airplane with an air speed of 130 mi/h has a heading of 200°. Find the speed and the direction of a wind which makes the plane travel with a ground speed of 140 mi/h on a 205° course.

23. An airplane has a cruising speed of 200 mi/h in calm air. A 40-mi/h wind is blowing in the direction 222°. The pilot wishes to fly 100 mi north and then return to his starting point. (*a*) In what direction

---

* Ignore the curvature of the earth and assume only two-place accuracy. Get angles to the nearest degree. Get distances to the nearest multiple of 10 mi.

should he head the plane on the outgoing trip? (*b*) What should be the heading on the return trip? (*c*) Find the time (to the nearest minute) required to make the round trip.

24. An airplane heads north (direction 0.0°) with an air speed of 200 km/h. A wind speed of 40.0 km/h causes the plane to move in the direction 4.0°. Find the wind direction and the ground speed of the plane.

25. A wind of 50.0 km/h is blowing in the direction 340.0°. By heading due west (270.0°), a pilot achieves a ground speed of 300 km/h. Find the air speed and the course of the plane.

26. A vertical 10-meter flagpole standing on a hillside casts a 4.5-meter shadow straight *up* the slope. At the tip of the shadow, the angle subtended by the pole is 78°. Find the angle made by the hillside with the horizontal. Find the elevation of the sun.

27. A pole stands on level ground and leans eastward, making an angle of 50° with the horizontal. The pole is supported by a 10-ft prop whose base is 12.8 ft from the base of the pole. Find the angle made by the prop with the horizontal.

28. A truck is moving at 70 mi/h on a straight road in the direction S 42° E. The sun, rising in the direction S 80° E, casts the truck's shadow onto the front of a building that faces due east. How fast is the truck's shadow moving?

29. A 50-ft flagpole stands on the top of a 20-ft building. How far from the base of the building should a person stand if the flagpole and the building are to subtend equal angles at her eye, which is 5 ft above the ground? (Consider the numbers as exact.)

30. A vertical tree stands on a hillside that makes an angle α with the horizontal. From a point directly *up* the hill from the tree, the angle of elevation of the treetop is β. From a point m meters farther up the hill, the angle of elevation of the treetop is γ. If the height of the tree is x meters, express x in terms of m, α, β, and γ.

31. A *segment* of a circle is the region bounded by an arc of the circle and the chord that subtends it. Show that the area of the shaded segment of the circle in Figure 84 is equal to $\frac{1}{2}r^2 (\theta - \sin \theta)$, where θ is in radians. (See Problem 23, Exercise 19.)

32. Given a, b, and A, where A is an acute angle, use the law of cosines to show that

$$c = b \cos A \pm \sqrt{a^2 - b^2 \sin^2 A}$$

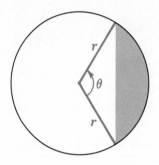

FIGURE 84

Discuss this equation and interpret the geometric significance if (1) $a < b \sin A$, (2) $a = b \sin A$, (3) $b \sin A < a < b$, (4) $a \geq b$. Compare with the possibilities in the *SSA* case.

33. The lengths of the sides of a triangle are 6, 7, and 10. Find the length of the median drawn to the longest side. Consider the numbers as exact. (A median of a triangle is the line segment connecting the vertex of an angle to the midpoint of the opposite side.)

34. Prove that in any triangle,

$$a^2 + b^2 + c^2 = 2ab \cos C + 2bc \cos A + 2ac \cos B$$

*35. Prove that in any triangle

$$\frac{a + b}{c} = \frac{\cos \tfrac{1}{2}(A - B)}{\sin \tfrac{1}{2}C}$$

Hint: Add the equations $\dfrac{a}{c} = \dfrac{\sin A}{\sin C}$ and $\dfrac{b}{c} = \dfrac{\sin B}{\sin C}$. Apply formula (15) of Section 49 to $\sin A + \sin B$. Replace $\sin C$ with $2 \sin \tfrac{1}{2}C \cos \tfrac{1}{2}C$. Notice that $\tfrac{1}{2}C$ is the complement of $\tfrac{1}{2}(A + B)$.

*36. Prove that in any triangle

$$\frac{a - b}{c} = \frac{\sin \tfrac{1}{2}(A - B)}{\cos \tfrac{1}{2}C}$$

Use cyclic permutation and the interchange of pairs of letters to derive five similar formulas.

---

*The formulas in Probs. 35 and 36 are called **Mollweide's equations.** They serve as good checks because each of them involves all six parts of the triangle.

12

INVERSE TRIGONOMETRIC FUNCTIONS

72 INVERSE TRIGONOMETRIC RELATIONS

(Before proceeding, the student should make a thorough review of Sections 34 and 35.) The equation

$$u = \sin \theta$$

says that u is a number representing the sine of the number* θ. Another interpretation is

$$\theta \text{ is a number* whose sine is } u$$

This statement is usually written in the form

$$\theta = \arcsin u$$

With this understanding, we can say

$$\arcsin \frac{1}{2} = \frac{\pi}{6}, \frac{5\pi}{6}, \frac{13\pi}{6}, -\frac{7\pi}{6}, \text{ etc.}$$

because the sine of each of these numbers is $\frac{1}{2}$. Values of $\arcsin \frac{1}{2}$ may be obtained by finding all the points on the sine curve that are $\frac{1}{2}$ unit above the θ axis (Figure 85).

---

*This number may be regarded as the radian measure of an angle.

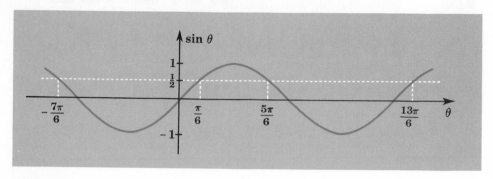

FIGURE 85

Similarly, arccos u denotes a number* whose cosine is u, arctan u denotes a number* whose tangent is u, etc. The six inverse trigonometric relations are arcsin u, arccos u, arctan u, arccot u, arcsec u, arccsc u. Thus arccos $0 = \pi/2$, $3\pi/2$, etc., and arctan $1 = \pi/4$, $5\pi/4$, etc.

73 INVERSE TRIGONOMETRIC FUNCTIONS

The sine function is defined as the infinite set of ordered pairs $\{(\theta, \sin\theta)$, where θ is a real number$\}$. This trigonometric relation† is a function because to each real number θ there corresponds one and only one number $\sin\theta$. Let us now consider the arcsin relation $\{(u, \text{arcsin } u)$, where $-1 \leq u \leq 1\}$. As demonstrated in the preceding section, to each value of u there corresponds more than one value of arcsin u. Therefore, the arcsin relation is not a function.

It will be quite useful to modify the arcsin relation in such a way as to make it a function. A careful inspection of the sine curve (Figure 86) shows that as θ varies from $-\pi/2$ to $\pi/2$, then $\sin\theta$ moves from -1 to $+1$, taking on all intervening values once and only once. Thus, if arcsin u is restricted to the closed interval $[-\pi/2, \pi/2]$, then the relation $\{(u, \text{arcsin } u)$, where $-\pi/2 \leq \text{arcsin } u \leq \pi/2\}$ is a function. The notation Arcsin u or $\text{Sin}^{-1} u$ (read "inverse sine u") means *the* number (between $-\pi/2$ and $\pi/2$, inclusive) whose sine is u; that is,

$$-\frac{\pi}{2} \leq \text{Sin}^{-1} u \leq \frac{\pi}{2}$$

Thus, $\text{Sin}^{-1}\frac{1}{2} = \frac{\pi}{6}$. It should be carefully noted that the superscript -1

---

*This number may be regarded as (the radian measure of) an angle.

† Recall that a relation is a set of ordered pairs of numbers, whereas a function is a relation $\{(x, y)\}$ such that to each x there corresponds one and only one y.

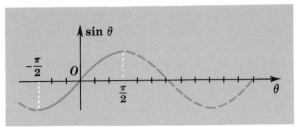

Find $\text{Sin}^{-1} u$ between $-\frac{\pi}{2}$ and $\frac{\pi}{2}$.

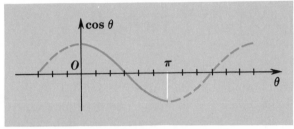

Find $\text{Cos}^{-1} u$ between 0 and π.

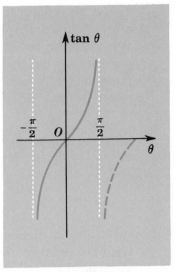

Find $\text{Tan}^{-1} u$ between $-\frac{\pi}{2}$ and $\frac{\pi}{2}$.

FIGURE 86

is *not* to be interpreted as an exponent; that is, $\text{Sin}^{-1} u$ is *not* $(\text{Sin } u)^{-1}$ or $\dfrac{1}{\sin u}$.

The symbols arccos u and arctan u may be modified to refer to functions rather than relations by defining Arccos $u = \text{Cos}^{-1} u$ and Arctan $u = \text{Tan}^{-1} u$ as follows:

$$0 \leq \text{Cos}^{-1} u \leq \pi \qquad -\frac{\pi}{2} < \text{Tan}^{-1} u < \frac{\pi}{2}$$

The following summary shows the domain, range, and defining equation for each of the previously discussed inverse trigonometric functions.

Function	Inverse function	Defining equation	Domain	Range
sin	Sin^{-1}	$y = \text{Sin}^{-1} x$	$-1 \leq x \leq 1$	$-\frac{\pi}{2} \leq \text{Sin}^{-1} x \leq \frac{\pi}{2}$
cos	Cos^{-1}	$y = \text{Cos}^{-1} x$	$-1 \leq x \leq 1$	$0 \leq \text{Cos}^{-1} x \leq \pi$
tan	Tan^{-1}	$y = \text{Tan}^{-1} x$	$-\infty < x < \infty$	$-\frac{\pi}{2} < \text{Tan}^{-1} x < \frac{\pi}{2}$

Figure 86 restates these definitions in terms of the sine, cosine, and tangent curves; the unbroken lines indicate the portions of the curves that are to be used.

Hence, by definition, $Sin^{-1} u$, $Cos^{-1} u$, and $Tan^{-1} u$ refer to functions rather than relations. Thus $Cos^{-1} u$ means *the* number (between 0 and π) whose cosine is u.

ILLUSTRATIONS $\quad Sin^{-1}\left(-\dfrac{\sqrt{3}}{2}\right) = -\dfrac{\pi}{3}$

$$Cos^{-1}\left(-\dfrac{\sqrt{2}}{2}\right) = \dfrac{3\pi}{4} \qquad Tan^{-1}(-1) = -\dfrac{\pi}{4}$$

In calculus, frequent use is made of the inverse trigonometric functions. For example, the area bounded by the curve $y = 1/(x^2 + 1)$, the x axis, and the lines $x = 0$ and $x = 2$ is $Tan^{-1} 2 = 1.1071$ (square units).

The functions $Cot^{-1} u$, $Sec^{-1} u$, and $Csc^{-1} u$ are of less importance. They can readily be expressed in terms of $Tan^{-1} u$, $Cos^{-1} u$, and $Sin^{-1} u$, respectively. For example, $Sec^{-1} u = Cos^{-1}(1/u)$ [the number whose secant is u is the same number whose cosine is $(1/u)$]. Thus, $Sec^{-1} 2 = Cos^{-1}\dfrac{1}{2} = \dfrac{\pi}{3}$.

EXAMPLE   Evaluate to four decimal places: (*a*) $Sin^{-1}(-0.8763)$, (*b*) $Cos^{-1}(-0.4321)$, (*c*) $Cot^{-1} 3.271$.

CALCULATOR SOLUTION   Most scientific (slide rule) calculators have a Change Sign key $\boxed{+/-}$ which changes the sign of the displayed number. Switch the calculator to radian mode.

(*a*) Press $\boxed{\cdot}\boxed{8}\boxed{7}\boxed{6}\boxed{3}\boxed{+/-}\boxed{inv}\boxed{sin}$

The displayed result is -1.068127636. Hence

$$Sin^{-1}(-0.8763) = -1.0681$$

This is the radian measure of a negative fourth-quadrant angle $(-61.2°)$.

(*b*) Press $\boxed{.4321}\boxed{+/-}\boxed{inv}\boxed{cos}$

The displayed result is 2.017616418. Therefore

$$\text{Cos}^{-1}(-0.4321) = 2.0176$$

This is the radian measure of a positive second-quadrant angle (115.6°).

(c) Since

$$\text{Cot}^{-1} 3.271 = \text{Tan}^{-1} \frac{1}{3.271}$$

press | 3.271 | | 1/x | | inv | | tan |

The displayed result is 0.296693375. Hence

$$\text{Cot}^{-1} 3.271 = 0.2967$$

TABLE SOLUTION   If the value of a trigonometric function of a first-quadrant angle is given, we can use Table 1 to find the radian measure of the angle.

(a) From Table 1, we find

$$\text{Sin}^{-1} 0.8763 = 1.0681$$

Therefore

$$\text{Sin}^{-1}(-0.8763) = -1.0681$$

(b) Table 1 gives us $\text{Cos}^{-1} 0.4321 = 1.1240$.

Hence

$$\begin{aligned} \text{Cos}^{-1}(-0.4321) &= \pi - 1.1240 \\ &= 3.1416 - 1.1240 \\ &= 2.0176 \end{aligned}$$

Observe that the second-quadrant angle 2.0176 has 1.1240 for its related angle.

(c) Since Table 1 contains cotangent values, we find

$$\text{Cot}^{-1} 3.271 = 0.2967$$

EXERCISE 36

Find the value of each of the following. Leave the results in terms of π. Do not use a calculator or tables.

1. $\text{Cos}^{-1} 1$

2. $\text{Cos}^{-1}\left(-\dfrac{\sqrt{3}}{2}\right)$

3. $\text{Sin}^{-1}(-1)$

4. $\text{Sin}^{-1}\dfrac{\sqrt{2}}{2}$

5. $\text{Sin}^{-1}\left(-\dfrac{\sqrt{2}}{2}\right)$

6. $\text{Sin}^{-1} 1$

7. $\text{Cos}^{-1}\dfrac{\sqrt{2}}{2}$

8. $\text{Cos}^{-1} 0$

9. $\text{Cos}^{-1}\dfrac{\sqrt{3}}{2}$

10. $\text{Cos}^{-1}\frac{1}{2}$

11. $\text{Sin}^{-1}\dfrac{\sqrt{3}}{2}$

12. $\text{Sin}^{-1}\left(-\frac{1}{2}\right)$

13. $\text{Sin}^{-1} 0$

14. $\text{Csc}^{-1} 2$

15. $\text{Cos}^{-1}\left(-\frac{1}{2}\right)$

16. $\text{Cos}^{-1}(-1)$

17. $\text{Tan}^{-1}(-\sqrt{3})$

18. $\text{Tan}^{-1} 1$

19. $\text{Tan}^{-1}\left(-\dfrac{1}{\sqrt{3}}\right)$

20. $\text{Cot}^{-1}\left(\dfrac{1}{\sqrt{3}}\right)$

21. $\text{Sec}^{-1}\sqrt{2}$

Use a calculator or Table 1 to find the value of each of the following to four decimal places:

22. $\text{Sin}^{-1}(-0.3811)$

23. $\text{Cos}^{-1} 0.1908$

24. $\text{Tan}^{-1}(-0.2530)$

25. $\text{Cos}^{-1}(-0.7912)$

26. $\text{Tan}^{-1} 8.643$

27. $\text{Sin}^{-1} 0.9252$

28. Explain why $\text{Cos}^{-1} u$ could not be taken on the interval $-\pi/2$ to $\pi/2$.

74 OPERATIONS INVOLVING INVERSE TRIGONOMETRIC FUNCTIONS

Since every inverse trigonometric function may be thought of as (the radian measure of) an angle, it is frequently convenient to place this angle in standard position and label its triangle of reference in accord-

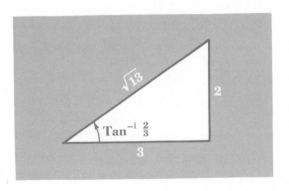

FIGURE 87

ance with the inverse function (Example 1). Sometimes it is advisable to replace the inverse functions with angle symbols, such as θ, A, B, and then try to express the problem in terms of ordinary functions.

EXAMPLE 1 Evaluate $\cos\left(\text{Tan}^{-1}\frac{2}{3}\right)$.

SOLUTION We are asked to find the cosine of the angle whose tangent is $\frac{2}{3}$. Draw a right triangle with legs 2 and 3. The acute angle opposite the side 2 has a tangent of $\frac{2}{3}$. It can be labeled $\text{Tan}^{-1}\frac{2}{3}$. After finding that the hypotenuse is $\sqrt{13}$, we see (Figure 87) that

$$\cos\left(\text{Tan}^{-1}\frac{2}{3}\right) = \frac{3}{\sqrt{13}}$$

EXAMPLE 2 Find the value of

$$\sin\left(\text{Sin}^{-1}u + \text{Cos}^{-1}v\right)$$

SOLUTION Since $\text{Sin}^{-1}u$ and $\text{Cos}^{-1}v$ are angles, we have the sine of the sum of two angles. Let $A = \text{Sin}^{-1}u$ and $B = \text{Cos}^{-1}v$. Using

$$\sin(A + B) = \sin A \cos B + \cos A \sin B$$

we have

$$\sin\left(\text{Sin}^{-1}u + \text{Cos}^{-1}v\right) = \sin\left(\text{Sin}^{-1}u\right)\cos\left(\text{Cos}^{-1}v\right)$$
$$+ \cos\left(\text{Sin}^{-1}u\right)\sin\left(\text{Cos}^{-1}v\right)$$
$$= uv + \sqrt{1 - u^2}\sqrt{1 - v^2}$$

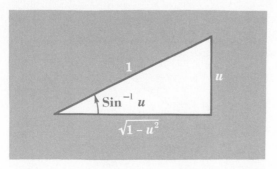

FIGURE 88

The value of $\cos(\text{Sin}^{-1} u)$ is found by use of the triangle in Figure 88. Notice that if u is negative, $\text{Sin}^{-1} u$ is in Q IV (its value in radians lies between $-\pi/2$ and 0) but $\cos(\text{Sin}^{-1} u)$ is still positive. Explain why $\sin(\text{Cos}^{-1} v)$ is always the positive radical $\sqrt{1 - v^2}$.

EXAMPLE 3 Prove that

$$\text{Tan}^{-1}\frac{1}{2} + \text{Tan}^{-1}\frac{1}{3} = \frac{\pi}{4}$$

SOLUTION The left side is the sum of two acute angles, each less than $\pi/4$. Why? To prove that their sum is $\pi/4$, let us take the tangent of each side of the equation:

$$\tan(\text{Tan}^{-1}\tfrac{1}{2} + \text{Tan}^{-1}\tfrac{1}{3}) \qquad\qquad \tan\frac{\pi}{4}$$

$$= \frac{\tan(\text{Tan}^{-1}\tfrac{1}{2}) + \tan(\text{Tan}^{-1}\tfrac{1}{3})}{1 - \tan(\text{Tan}^{-1}\tfrac{1}{2})\tan(\text{Tan}^{-1}\tfrac{1}{3})} \qquad = 1$$

$$= \frac{\tfrac{1}{2} + \tfrac{1}{3}}{1 - \tfrac{1}{2}\cdot\tfrac{1}{3}} = \frac{\tfrac{5}{6}}{\tfrac{5}{6}} = 1$$

The proof is complete if we recall that two acute angles having the same tangent are equal. Notice that the formula for $\tan(A + B)$ was used in evaluating the left side.

EXAMPLE 4 Solve for x: $\text{Sin}^{-1} 2x + \text{Cos}^{-1} x = \dfrac{\pi}{6}$

SOLUTION Subtract $\text{Cos}^{-1} x$ from both sides of the equation. Then take the sine of both sides of the equation:

$$\sin (\text{Sin}^{-1} 2x) = \sin \left(\frac{\pi}{6} - \text{Cos}^{-1} x \right)$$

Let $A = \frac{\pi}{6}$ and $B = \text{Cos}^{-1} x$. Using

$$\sin (A - B) = \sin A \cos B - \cos A \sin B$$

we obtain

$$2x = \sin \frac{\pi}{6} \cos (\text{Cos}^{-1} x) - \cos \frac{\pi}{6} \sin (\text{Cos}^{-1} x)$$

$$2x = \frac{1}{2} \cdot x - \frac{\sqrt{3}}{2} \cdot \sqrt{1 - x^2}$$

$$3x = - \sqrt{3} \cdot \sqrt{1 - x^2}$$

Squaring both sides, we obtain

$$9x^2 = 3 - 3x^2 \qquad 4x^2 = 1 \qquad x = \pm \tfrac{1}{2}$$

Inasmuch as we squared the equation, we must check all values to see if any are extraneous.

Check for $x = \tfrac{1}{2}$:

$$\text{Sin}^{-1} 1 + \text{Cos}^{-1} \frac{1}{2} = \frac{\pi}{6}$$

$$\frac{\pi}{2} + \frac{\pi}{3} = \frac{\pi}{6} \qquad \text{False}$$

Check for $x = -\tfrac{1}{2}$:

$$\text{Sin}^{-1} (-1) + \text{Cos}^{-1} \left(-\frac{1}{2} \right) = \frac{\pi}{6}$$

$$-\frac{\pi}{2} + \frac{2\pi}{3} = \frac{\pi}{6} \qquad \text{True}$$

Hence the only solution of the given equation is $x = -\tfrac{1}{2}$.

$$\overline{\text{EXERCISE 37}}$$

Find the exact value of each of the following. Do not use a calculator or tables.

1. csc (Sin^{-1}u)

2. sec (Tan^{-1}u)

3. tan (Sin^{-1}u)

4. csc (Csc^{-1}u)

5. cos [Tan$^{-1}(-\frac{5}{12})$]

6. sin (Sec$^{-1}6$)

7. tan [Sin$^{-1}(-\frac{4}{5})$]

8. tan [Cos$^{-1}(-\frac{1}{4})$]

9. sin (Cos^{-1}u − Cos^{-1}v)

10. cos (Cos^{-1}u − Sin^{-1}v)

11. cos (Sin^{-1}u + Cos^{-1}v)

12. sin (Sin^{-1}u + Sin^{-1}v)

13. $\cos\left(\dfrac{\pi}{6} + \text{Sin}^{-1}\dfrac{5}{8}\right)$

14. sin (π + Sin$^{-1}\frac{1}{3}$)

15. $\sin\left(\dfrac{\pi}{4} - \text{Cos}^{-1}v\right)$

16. $\cos\left(\dfrac{3\pi}{2} - \text{Cos}^{-1}\dfrac{5}{7}\right)$

17. Sin (2 Cos^{-1}u)

Hint: Let $A = $ Cos^{-1}u. We seek sin 2A.

18. cos (2 Sin^{-1}u)

19. cos (2 Cos^{-1}v)

20. sin (2 Sin^{-1}v)

21. cos (Cos$^{-1}2v$)

22. sin (Sin$^{-1}4u$)

23. $\text{Sin}^{-1}\left(\sin\dfrac{4\pi}{3}\right)$

24. $\text{Cos}^{-1}\left[\cos\left(-\dfrac{4\pi}{9}\right)\right]$

25. $\text{Sin}^{-1}\left(\sin\dfrac{3\pi}{5}\right)$

26. $\text{Tan}^{-1}\left(\tan\dfrac{6\pi}{7}\right)$

27. $\text{Tan}^{-1}\left(\cot\dfrac{\pi}{5}\right)$

28. Tan^{-1} ($-\sqrt{4 + \cos\pi}$)

29. $\text{Cos}^{-1}\sqrt{1 - \sin^4\dfrac{7\pi}{4}}$

30. $\text{Sin}^{-1}\left[\tan\left(-\dfrac{\pi}{4}\right)\right]$

Assume u > 0. Copy the following and fill in the blanks:

31. $\text{Tan}^{-1}\dfrac{u}{\sqrt{1 - u^2}} = \text{Sin}^{-1}\underline{\hspace{1cm}} = \text{Cos}^{-1}\underline{\hspace{1cm}}, \; 0 < u < 1$

32. $\text{Sin}^{-1}\dfrac{\sqrt{16 - u^2}}{4} = \text{Cos}^{-1}\underline{\hspace{1cm}} = \text{Tan}^{-1}\underline{\hspace{1cm}}, \; 0 < u < 4$

33. $\text{Cos}^{-1}\dfrac{2}{\sqrt{u^2 + 2u + 5}} = \text{Sin}^{-1}\underline{\hspace{1cm}} = \text{Tan}^{-1}\underline{\hspace{1cm}}$

34. $\text{Csc}^{-1}\dfrac{u}{\sqrt{u^2 - 36}} = \text{Cot}^{-1}\underline{\hspace{1cm}} = \text{Sec}^{-1}\underline{\hspace{1cm}}, \; u > 6$

Prove each of the following identities without using a calculator or tables. (Consider all permissible values—positive, negative, and zero—of the letters involved.)

35. $\cos\left(\frac{1}{2}\operatorname{Cos}^{-1}u\right) = \sqrt{\frac{1+u}{2}}$

36. $\sin\left(\frac{1}{2}\operatorname{Cos}^{-1}3a\right) = \sqrt{\frac{1-3a}{2}}$

37. $\sin\left(\frac{1}{2}\operatorname{Sin}^{-1}c\right) = \pm\sqrt{\frac{1-\sqrt{1-c^2}}{2}}$

38. $\cos\left(\frac{1}{2}\operatorname{Sin}^{-1}2u\right) = \sqrt{\frac{1+\sqrt{1-4u^2}}{2}}$

39. $\operatorname{Tan}^{-1}9 - \operatorname{Tan}^{-1}7 = \operatorname{Tan}^{-1}\frac{1}{32}$

40. $\operatorname{Cos}^{-1}\frac{1}{7} - \operatorname{Cos}^{-1}\frac{13}{14} = \frac{\pi}{3}$

41. $\operatorname{Sin}^{-1}\frac{12}{13} - \operatorname{Sin}^{-1}\frac{4}{5} = \operatorname{Sin}^{-1}\frac{16}{65}$

42. $2\operatorname{Cos}^{-1}\frac{1}{3} + \operatorname{Sin}^{-1}\left(-\frac{7}{9}\right) = \frac{\pi}{2}$

43. $\frac{1}{2}\operatorname{Cos}^{-1}\left(-\frac{7}{8}\right) = \operatorname{Sin}^{-1}\frac{7}{8} + \operatorname{Sin}^{-1}\frac{1}{4}$

44. $\operatorname{Tan}^{-1}1 + \operatorname{Tan}^{-1}2 + \operatorname{Tan}^{-1}3 = \pi$

Identify each statement as true or false and give reasons. (Consider all permissible values—positive, negative, and zero—of the letters involved.)

45. $\operatorname{Sin}^{-1}(-u) = -\operatorname{Sin}^{-1}u$

46. $\operatorname{Cos}^{-1}(-u) = \pi - \operatorname{Cos}^{-1}u$

47. $\operatorname{Tan}^{-1}(-u) = -\operatorname{Tan}^{-1}u$

48. $\csc(\operatorname{Cos}^{-1}a) = \frac{1}{a}$

49. $\operatorname{Tan}^{-1}(+\infty) = \frac{\pi}{2}$

50. $\operatorname{Sin}^{-1}u + \operatorname{Cos}^{-1}u = \frac{\pi}{2}$

51. $\cos(2\operatorname{Sin}^{-1}u) = 1 - 2u^2$

52. $\sin(\operatorname{Cos}^{-1}u) = \cos(\operatorname{Sin}^{-1}u)$

53. $\operatorname{Sin}^{-1}u = \operatorname{Cos}^{-1}\sqrt{1-u^2}$

54. $\operatorname{Tan}^{-1}u + \operatorname{Tan}^{-1}\frac{1}{u} = \frac{\pi}{2}$

55. $\sec\left(\mathrm{Sin}^{-1} u\right) = \pm\dfrac{1}{\sqrt{1 - u^2}}$

56. $\mathrm{Cot}^{-1} u = \dfrac{1}{\mathrm{Tan}^{-1} u}$

Solve for x.

57. $\mathrm{Sin}^{-1}\sqrt{2x} = \mathrm{Cos}^{-1}\sqrt{x}$

58. $\mathrm{Cos}^{-1} 2x + \mathrm{Sin}^{-1} x = \dfrac{5\pi}{6}$

59. $\mathrm{Cos}^{-1} x + \mathrm{Tan}^{-1} x = \dfrac{\pi}{2}$

60. $\mathrm{Sin}^{-1} 2x = \mathrm{Cos}^{-1} x$

61. $\mathrm{Cos}^{-1}(3x - 2) = 2\,\mathrm{Cos}^{-1} x$

62. $2\,\mathrm{Sin}^{-1} x + \mathrm{Cos}^{-1} x = \pi$

63. $\mathrm{Cos}^{-1}\left(\dfrac{x}{\sqrt{3}}\right) = \pi - \mathrm{Sin}^{-1}(-x)$

64. Given the equation $\mathrm{Tan}^{-1} u = \dfrac{1}{2}\,\mathrm{Sin}^{-1}\dfrac{2u}{1 + u^2}$.

(*a*) Prove that this equation is, or is not, an identity for $0 \leq u \leq 1$.
(*b*) Prove that the equation does not hold for any value of $u > 1$.

75 INVERSE FUNCTIONS*

Two equations are equivalent if they have the same solution set. That is, every solution of one equation is a solution of the other, and vice versa. Thus $y = 2^x$ and $x = \log_2 y$ are equivalent equations by the definition of logarithm. If f and g are functions and if $y = f(x)$ and $x = g(y)$ are equivalent equations, then f and g are said to be inverses of each other. To find the inverse of the function f, where $f(u) = 2u + 7$, we let $y = 2x + 7$† and then solve for x: $x = \frac{1}{2}(y - 7)$. Hence the inverse of f is g, where $g(u) = \frac{1}{2}(u - 7)$. Likewise, if $y = \sin x$, where $-\pi/2 \leq x \leq \pi/2$, then $x = \mathrm{Sin}^{-1}y$, where $-1 \leq y \leq 1$. Hence, if $-\pi/2 \leq u \leq \pi/2$, then the inverse of the trigonometric function

---

*For a detailed treatment of inverse functions, see the author's *College Algebra*, McGraw-Hill, New York, 1973.

† The quantity $2x + 7$ is a function of x. [This is another way of describing the function consisting of all ordered pairs $(x, 2x + 7)$, for real values of x.] The expression $2u + 7$ is the *same* function of u. In each case the independent variable is doubled and added to 7.

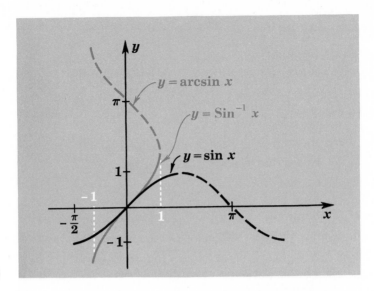

FIGURE 89

$\{(u, \sin u)\}$ is the inverse trigonometric function $\{(u, \operatorname{Sin}^{-1} u)\}$. For this reason Sin^{-1} is called an inverse trigonometric function.

If f and g are inverse functions, then it can be shown that $g[f(u)] = u$ and $f[g(u)] = u$. If $f(u) = 2^u$, then $g(u) = \log_2 u$ (see Section 62). Thus $g[f(u)] = \log_2 (2^u) = u$ and $f[g(u)] = 2^{\log_2 u} = u$. Moreover, if $f(u) = \sin u$, where $-\pi/2 \le u \le \pi/2$, then $f[g(u)] = \sin (\operatorname{Sin}^{-1} u) = u$ and $g[f(u)] = \operatorname{Sin}^{-1} (\sin u) = u$.

Furthermore, if the function g^* is the inverse of the function f, then the graph of $g(u)$ is the reflection of the graph of $f(u)$ in the line that bisects the first and third quadrants. (See Section 62.) Hence the graph of $y = \operatorname{Sin}^{-1} x$ is the reflection of the graph of $y = \sin x$ (for $-\pi/2 \le x \le \pi/2$) in the line $y = x$ (Figure 89).

EXERCISE 38

In the following problems, g designates the inverse of the function f.

1. Given $f(u) = \dfrac{2u + 3}{4}$.

    (*a*) Find $g(u)$.

    (*b*) Show that $g[f(u)] = f[g(u)] = u$.

---

*It would be consistent to use f^{-1} to designate the inverse of f. Then $g[f(u)] = f^{-1}f^1(u) = f^0(u) = u$, *symbolically.* Of course we are not multiplying (in the narrow sense) f^{-1} by f.

2. Given $f(u) = \dfrac{4u - 5}{u - 6}$ $u \neq 6$.

   (*a*) Find $g(u)$.

   (*b*) Show that $g[f(u)] = u$, provided $u \neq 6$.

   (*c*) Show that $f[g(u)] = u$, provided $u \neq 4$.

3. Given $f(u) = 3 + \sqrt{u - 2}$ $u \geq 2$.

   (*a*) Find $g(u)$.

   (*b*) Show that $g[f(u)] = u$, provided $u \geq 2$.

   (*c*) Show that $f[g(u)] = u$, provided $u \geq 3$.

4. Given $f(u) = 1 + \sqrt[3]{u}$.

   (*a*) Find $g(u)$.

   (*b*) Show that $g[f(u)] = f[g(u)] = u$.

5. Sketch, on the same coordinate system, the graphs of $y = \cos x$ and $y = \text{Cos}^{-1} x$.

6. Sketch, on the same coordinate system, the graphs of $y = \tan x$ and $y = \text{Tan}^{-1} x$.

13 COMPLEX NUMBERS

76 COMPLEX NUMBERS

It will be recalled that there are quadratic equations with real coefficients that have no real roots. The simplest such equation is $x^2 + 1 = 0$, which obviously has no real solution, since the square of any real number—whether positive, negative, or zero—is nonnegative. (An analogous linear equation with positive coefficients is $x + 1 = 0$, which necessitates the postulation of negative numbers for its solution—though the negative numbers probably arose from other considerations.) Another example of a quadratic equation with real coefficients but with no real roots is $x^2 - 4x + 13 = 0$. It was for the solution of such equations that the system of *complex numbers* was first investigated, but this remarkable system has proved to be invaluable in many other connections.

The system of complex numbers is really the system of *ordered pairs* of *real* numbers—a first, then a second—(a, b), in which equality, addition, and multiplication are defined in a certain specified way. Usually, however, complex numbers are written as $a + bi$, and in this form the defining properties* are as follows:

> Equality: $a + bi = c + di$ if and only if $a = c$ and $b = d$
>
> Addition: $(a + bi) + (c + di) = (a + c) + (b + d)i$
>
> Multiplication: $(a + bi)(c + di) = (ac - bd) + (ad + bc)i$

---

*In terms of the (a, b) notation, which we shall not adopt, these definitions are
Equality: $(a, b) = (c, d)$ if and only if $a = c$ and $b = d$
Addition: $(a, b) + (c, d) = (a + c, b + d)$
Multiplication: $(a, b)(c, d) = (ac - bd, ad + bc)$

It is easy to show that all the ordinary rules for adding and multiplying real numbers—the associative and commutative laws, etc.—carry over to the system of complex numbers. For this reason, we speak of the *field* of complex numbers.

There is no need to memorize the definitions of addition and multiplication of complex numbers. Just remember that addition and multiplication are ordinary addition and multiplication, with the single special property that

$$i^2 = -1$$

There is, however, one property of the real numbers that does not carry over to the complex numbers. *The complex number field is not ordered;* we do not say that one complex number is less than or greater than another.

Complex numbers have been defined as numbers that obey the definitions of equality, addition, and multiplication listed above. This approach is equivalent to the following definition.

> A *complex number* is a number of the form $a + bi$, where a and b are real numbers and $i = \sqrt{-1}$.
>
> If $b \neq 0$, the complex number $a + bi$ is called an *imaginary number*.
>
> If $a = 0$ and $b \neq 0$, the complex number $a + bi$ is called a *pure imaginary number*.
>
> If $b = 0$, the complex number $a + bi$ becomes a, a real number.

Hence we see that the field of complex numbers includes all real numbers and all imaginary numbers.

ILLUSTRATIONS

Imaginary numbers:	$3i$, $2 + 5i$, $-7 + 8i$, $9 - i$, $-1 - 6i$
Pure imaginary numbers:	$3i$, $-4i$, $\sqrt{-49}$, $-\sqrt{-2}$
Real numbers:	4, $\frac{1}{7}$, 0, $-\frac{2}{9}$, 5, $-\sqrt{2}$, π

All these numbers are complex numbers.

Since $\sqrt{a} \cdot \sqrt{a} = a$ (by the definition of square root), we see that if $i = \sqrt{-1}$, then $i^2 = -1$. Moreover, $i^3 = i^2 \cdot i = -i$, $i^4 = (i^2)^2 = (-1)^2 = 1, i^5 = i^4 \cdot i = i, i^6 = i^4 \cdot i^2 = -1, i^{87} = i^{84} \cdot i^3 = (i^4)^{21} i^3 = 1^{21}(-i) = -i$.

In solving the equation $x^2 - 4x + 13 = 0$, we find the roots to be $x = 2 \pm \sqrt{-9}$. Remembering that $i = \sqrt{-1}$, we have $x = 2 \pm 3i$. Notice that if $2 + 3i$ is substituted for x in the equation $x^2 - 4x + 13 = 0$, we get

$$4 + 12i + 9i^2 - 8 - 12i + 13 = 0$$

If i^2 is replaced by -1, we obtain $17 + 12i - 12i - 17 = 0$. This shows that $2 + 3i$ is a perfectly good root of the equation, provided we understand that i is a number whose square is -1.

The complex numbers $a + bi$ and $a - bi$ are said to be *complex conjugates* of each other. Notice that the roots of $x^2 + 4 = 0$ are $2i$ and $-2i$, which are pure imaginary complex conjugates. The roots of the equation $x^2 - 4x + 13 = 0$ are the conjugate imaginary numbers $2 + 3i$ and $2 - 3i$. It can be shown that if an imaginary number $(a + bi)$ is a root of an equation with real coefficients, then the conjugate imaginary $(a - bi)$ is also a root of this equation.

Since i is a number whose square is -1, the best procedure in handling complex numbers is to perform all operations as if i were an ordinary letter and then replace i^2 with -1. It is to be noted that the quotient of two complex numbers is obtained by multiplying numerator and denominator by the conjugate of the denominator. For example,

$$\frac{7 + 5i}{3 - i} = \frac{(7 + 5i)(3 + i)}{(3 - i)(3 + i)} = \frac{21 + 7i + 15i + 5i^2}{9 - i^2}$$

$$= \frac{16 + 22i}{10} = \frac{8}{5} + \frac{11}{5}i$$

This result can be checked by multiplying $\frac{8}{5} + \frac{11}{5}i$ by $3 - i$. What should the result be?

All complex numbers should first be written in the form $a + bi$. Thus $3 + \sqrt{-49} = 3 + \sqrt{49}\sqrt{-1} = 3 + 7i$. This procedure is suggested to avoid mistakes such as $\sqrt{-5} \cdot \sqrt{-5} = \sqrt{(-5)(-5)} = \sqrt{25} = 5$. This is obviously incorrect because, by the definition of square root, $\sqrt{-5}$ is a number which when multiplied by itself becomes -5. The correct way of handling this is $\sqrt{-5} \cdot \sqrt{-5} = i\sqrt{5} \cdot i\sqrt{5} = 5i^2 = -5$. This result agrees with the definition of square root.

EXERCISE 39

Perform each of the indicated operations and express the result in the form a + bi.

1. $(5 - 2i) + (-1 + 9i)$
2. $(-6 + 4i) - (3 + 7i)$
3. $(7 - i) + (-4 + 6i) - (9 + 8i)$
4. $(3 + 2i) - (8 - 5i)$
5. $(8 + 3i)(-7 + i)$
6. $(2 + i\sqrt{3})(4 - i\sqrt{3})$
7. $(6 - \sqrt{-7})(5 - \sqrt{-7})$
8. $(1 - 9i)(7 - 4i)$
9. $(9 - 6i)^2$
10. $(1 + i)^2$
11. i^{39}
12. $i^{100} + i^{102}$
13. $i^{21} + i^{22} + i^{23}$
14. i^{1980}
15. $\dfrac{7 - 2i}{4 - i}$
16. $\dfrac{8 - 5i}{6 - 7i}$
17. $\dfrac{5 - i\sqrt{2}}{3 + i\sqrt{2}}$
18. $\dfrac{-3 + 28i}{2 + 3i}$
19. $\dfrac{4i}{1 + i}$
20. $\dfrac{9 + 10i}{i}$

Find the values of the real numbers x and y.

21. $5x + 8yi = 15 - 32i$
22. $(x + 2i)(3 - 5i) = 25 + yi$
23. $(1 + xi)(y + i) = 2 + 49i$
24. $(x + 4i)(6 - i) = y + 24i$

25. Find the value of $4x^2 + 4x + 5$ if $x = -\frac{1}{2} + i$.
26. Show by substitution that $\frac{2}{3} - 4i$ is a root of the equation $9x^2 - 12x + 148 = 0$.
27. Write a complex number that is not an imaginary number.
28. Prove that the sum of two conjugate imaginary numbers is a real number.
29. Prove that the product of two conjugate imaginary numbers is a positive real number.

77 GRAPHICAL REPRESENTATION OF COMPLEX NUMBERS

Let us represent (as in the case of the x axis of a rectangular coordinate system) the real numbers by points on a horizontal directed line (Figure 90). Let the vector V represent the directed segment connecting the

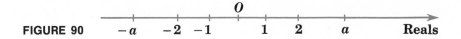

FIGURE 90

origin O to the point corresponding to the real number a. Since $ai^2 = -a$; it can be said that multiplying a by $i \cdot i$ is geometrically equivalent to rotating V through $180°$ about O. Consequently, it is logical to represent the multiplication of a by i as a rotation of V through $90°$ about O. Accordingly, the number ai will be represented as a point a units above O on the *vertical* line through O. We shall refer to the horizontal axis as the **axis of reals** and the vertical axis as the **axis of (pure) imaginaries.** This system of axes defines a region called the **complex plane.** It is to be noted that, while the unit on the axis of reals is the number 1, the unit on the axis of imaginaries is the imaginary number i. Hence the complex number $(a + bi)$ is represented by the point a units from the axis of imaginaries and b units from the axis of reals. Figure 91 illustrates the graphical representation of complex numbers in the complex plane.

It is often convenient to think of the complex number $(a + bi)$ as representing the vector OP (Figure 91).

78 GRAPHICAL ADDITION OF COMPLEX NUMBERS

Since the sum of $(a + bi)$ and $(c + di)$ is $(a + c) + (b + d)i$, we can add the numbers graphically by adding the real parts, a and c, to get the real part of the sum, and adding the imaginary coefficients, b and d, to get the

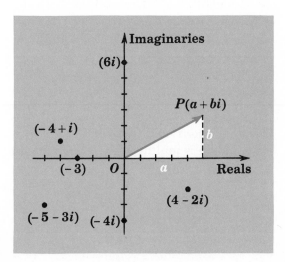

FIGURE 91

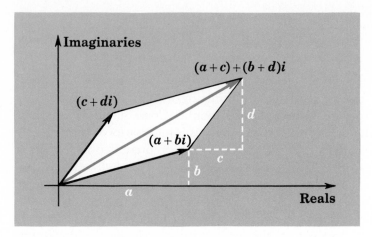

FIGURE 92

imaginary coefficient. This is illustrated in Figure 92. The result is exactly the same as if we had applied the parallelogram law to the vectors representing the numbers $(a + bi)$ and $(c + di)$. Three complex numbers can be added graphically by first obtaining the sum of two of them and then adding this to the third.

We can subtract $(c + di)$ from $(a + bi)$ graphically by adding $(a + bi)$ to $(-c - di)$.

79 TRIGONOMETRIC FORM OF A COMPLEX NUMBER

Let point P in the complex plane represent the complex number $a + bi$. The **absolute value*** of $a + bi$ is the distance r from O to P. It is always considered positive. The **amplitude*** of $a + bi$ is the angle measured from the positive axis of reals to the line OP. From Figure 93, it is obvious that

$$r = \sqrt{a^2 + b^2} \qquad \tan \theta = \frac{b}{a} \qquad (1)$$

and $$a = r \cos \theta \qquad b = r \sin \theta \qquad (2)$$

These equations hold regardless of the quadrant in which P lies. If the last equation is multiplied by i and added to the preceding one, we get

$$a + bi = r(\cos \theta + i \sin \theta) \qquad (3)$$

---

*Absolute value is also called *modulus*; amplitude is sometimes called *argument*.

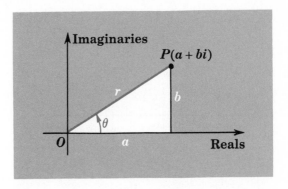

FIGURE 93

The expression $r(\cos\theta + i\sin\theta)$ is called the **trigonometric*** form of a complex number. The expression $a + bi$ is called the **algebraic form** of a complex number. The trigonometric form is useful in finding powers and roots of complex numbers.

Any complex number in algebraic form can be expressed in trigonometric form by the use of Equations (1). After the value of $\tan\theta$ has been obtained, θ can be found from a table of trigonometric functions. In general, there are two angles between $0°$ and $360°$ that have the same tangent. In order to be certain to get the correct angle, we should *always plot the complex number†* in the complex plane. The amplitude of a real number or a pure imaginary number can be obtained by inspection of its location in the complex plane. For example, the amplitude of $-4i$ is $270°$ (Figure 91).

Any complex number in trigonometric form can be expressed in algebraic form by use of Equations (2).

EXAMPLE   Express each of the following in trigonometric form:
(a) $3 - 3i$ (b) -4

SOLUTION   (a) Plot the number in the complex plane. Equations (1) give us $r = \sqrt{18} = 3\sqrt{2}$, $\tan\theta = -3/3 = -1$. From the last equation, θ could be $135°$ or $315°$. From Figure 94 we see that θ must be $315°$. Hence

$$3 - 3i = 3\sqrt{2}\,(\cos 315° + i\sin 315°)$$

---

*Also called the *polar* form. It is sometimes written in the abbreviated form $r\,\mathrm{cis}\,\theta$.
† The expression *plot the complex number* is an abbreviation we shall use for the more rigorous statement, *plot the point corresponding to the complex number*.

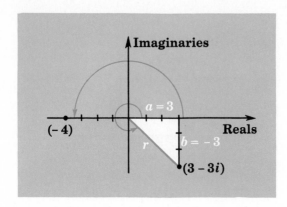

FIGURE 94

This result can be checked by replacing $\cos 315°$ and $\sin 315°$ with $\sqrt{2}/2$ and $-\sqrt{2}/2$, respectively, and then demonstrating that the right side is actually equal to the left side.

(b) After plotting the number, Figure 94, we find by inspection that $r = 4$ and $\theta = 180°$. Hence we can see immediately that

$$-4 = 4(\cos 180° + i \sin 180°)$$

It is to be carefully noted that, regardless of the signs of a and b, r is *always positive,* and *the signs in front of cos θ and i sin θ are always positive.*

EXERCISE 40

Perform the indicated operations graphically and check the results algebraically.

1. $(2 + 3i) + (-5 + i)$ **2.** $(1 - 6i) + (3 + 2i)$

3. $(-4 - 2i) - (1 + 5i)$ **4.** $(5 - 3i) + (-2 + 4i)$

5. $(6 - i) - (4 - 5i)$ **6.** $(-3 + i) - (-3 + 6i)$

7. $(7 - 4i) + (-6 + 2i) + 3i$

8. $(3 + 6i) + (2 - 5i) - (6 + i)$

Plot each of the following complex numbers and then express it in trigonometric form:

9. $2 - 2i$ **10.** $8 + 8i$ **11.** $-\sqrt{2} - i\sqrt{2}$

12. $-4 + 4i$ **13.** -9 **14.** $-6i$

15. 3   **16.** $7i$ **17.** $-2\sqrt{3} - 2i$

18. $4 - 3i$ **19.** $-5 - 12i$ **20.** $1 + i\sqrt{3}$

Plot each of the following complex numbers and then express it in algebraic form. Write the result with three-decimal-place accuracy.

21. $4(\cos 20° + i\sin 20°)$ **22.** $10(\cos 110° + i\sin 110°)$

23. $8(\cos 330° + i\sin 330°)$ **24.** $6(\cos 240° + i\sin 240°)$

25. On one system of coordinates, plot and label the number $2 + 3i$, its conjugate, and its negative.

26. What is the amplitude (a) of a positive real number? (b) of a negative real number? (c) of bi if $b > 0$? (d) of bi if $b < 0$?

27. Show that the negative of $r(\cos\theta + i\sin\theta)$ is

$$r[\cos(\theta + 180°) + i\sin(\theta + 180°)]$$

28. Show that the conjugate of $r(\cos\theta + i\sin\theta)$ is

$$r[\cos(-\theta) + i\sin(-\theta)]$$

80 MULTIPLICATION OF COMPLEX NUMBERS IN TRIGONOMETRIC FORM

THEOREM   The absolute value of the product of two complex numbers is the product of their absolute values; the amplitude of the product is the sum of their amplitudes:

$$r_1(\cos\theta_1 + i\sin\theta_1) \cdot r_2(\cos\theta_2 + i\sin\theta_2)$$
$$= r_1r_2[\cos(\theta_1 + \theta_2) + i\sin(\theta_1 + \theta_2)]$$

PROOF   Let $r_1(\cos\theta_1 + i\sin\theta_1)$ and $r_2(\cos\theta_2 + i\sin\theta_2)$ be any two complex numbers in trigonometric form. Their product is

$$r_1(\cos\theta_1 + i\sin\theta_1) \cdot r_2(\cos\theta_2 + i\sin\theta_2)$$
$$= r_1r_2(\cos\theta_1\cos\theta_2 + i\sin\theta_1\cos\theta_2 + i\cos\theta_1\sin\theta_2 + i^2\sin\theta_1\sin\theta_2)$$
$$= r_1r_2[(\cos\theta_1\cos\theta_2 - \sin\theta_1\sin\theta_2) + i(\sin\theta_1\cos\theta_2 + \cos\theta_1\sin\theta_2)]$$
$$= r_1r_2[\cos(\theta_1 + \theta_2) + i\sin(\theta_1 + \theta_2)]$$

ILLUSTRATION

$$2(\cos 130° + i \sin 130°) \cdot 3(\cos 50° + i \sin 50°)$$
$$= 2 \cdot 3[\cos (130° + 50°) + i \sin (130° + 50°)]$$
$$= 6(\cos 180° + i \sin 180°) = 6(-1 + i \cdot 0) = -6$$

This theorem can be extended to include the product of any number of complex numbers:

$$r_1(\cos \theta_1 + i \sin \theta_1) \cdot r_2(\cos \theta_2 + i \sin \theta_2) \cdots r_n(\cos \theta_n + i \sin \theta_n)$$
$$= r_1 r_2 \cdots r_n[\cos (\theta_1 + \theta_2 + \cdots + \theta_n) + i \sin (\theta_1 + \theta_2 + \cdots + \theta_n)]$$

81 DE MOIVRE'S THEOREM

If n is any real number,

$$[r(\cos \theta + i \sin \theta)]^n = r^n(\cos n\theta + i \sin n\theta)$$

PROOF   For n a positive integer (by mathematical induction)

Part 1   *Verification*

For $n = 1$:

$$[r(\cos \theta + i \sin \theta)]^1 = r(\cos \theta + i \sin \theta) \quad \text{True}$$

For $n = 2$:

$$[r(\cos \theta + i \sin \theta)]^2 = r^2(\cos 2\theta + i \sin 2\theta) \quad \text{True}$$

For $n = 3$:

$$[r(\cos \theta + i \sin \theta)]^3 = r^3(\cos 3\theta + i \sin 3\theta) \quad \text{True}$$

Part 2   A *proof* that *if* the theorem is true for $n = k$, then the theorem is true for $n = k + 1$. Let k represent any particular value of n. Assuming that

$$[r(\cos \theta + i \sin \theta)]^k = r^k(\cos k\theta + i \sin k\theta) \qquad (A)$$

we must prove that

$$[r(\cos\theta + i\sin\theta)]^{k+1}$$
$$= r^{k+1}[\cos(k+1)\theta + i\sin(k+1)\theta] \quad (B)$$

An examination of the left sides of the equations suggests that we multiply Equation (A) by $r(\cos\theta + i\sin\theta)$. Doing this, we obtain

$$[r(\cos\theta + i\sin\theta)]^{k+1}$$
$$= r^k(\cos k\theta + i\sin k\theta) \cdot r(\cos\theta + i\sin\theta)$$

Applying the theorem in Section 80, we get

$$[r(\cos\theta + i\sin\theta)]^{k+1} = r^{k+1}[\cos(k\theta + \theta) + i\sin(k\theta + \theta)]$$
$$= r^{k+1}[\cos(k+1)\theta + i\sin(k+1)\theta]$$

which is identical with Equation (B). This proves that if the theorem is true for $n = k$, then it must be true for $n = k + 1$.

Part 3 *Conclusion*   The theorem is true for $n = 1, 2, 3$ (Part 1). Since it is true for $n = 3$, it is true for $n = 4$ (Part 2, where $k = 3$ and $k + 1 = 4$). Since it is true for $n = 4$, it is true for $n = 5$, and so on for all positive integers n.*

It can be shown that De Moivre's theorem is true for all real values of n. We shall use it for only two cases: (1) when n is a positive integer, and (2) when n is the reciprocal of a positive integer. The proof of the latter case is omitted in this text.

EXAMPLE   Use De Moivre's theorem to find the value of $(-1 + i)^{10}$.

SOLUTION   After plotting $(-1 + i)$ and putting it in trigonometric form, we have

$$-1 + i = \sqrt{2}(\cos 135° + i\sin 135°)$$

---

*We could, of course, merely cite the axiom of mathematical induction as the authority for stating that De Moivre's theorem is true for all natural numbers n. The author feels, however, that the argument in Part 3 is helpful in establishing the plausibility of the axiom.

Apply De Moivre's theorem:

$$(-1 + i)^{10} = [\sqrt{2}(\cos 135° + i \sin 135°)]^{10}$$
$$= (\sqrt{2})^{10}(\cos 10 \cdot 135° + i \sin 10 \cdot 135°)$$
$$= 32(\cos 1350° + i \sin 1350°)$$
$$= 32(\cos 270° + i \sin 270°)$$
$$= -32i$$

82 ROOTS OF COMPLEX NUMBERS

THEOREM   The n nth roots of $r(\cos \theta + i \sin \theta)$ are given by the formula

$$\sqrt[n]{r}\left(\cos \frac{\theta + k \cdot 360°}{n} + i \sin \frac{\theta + k \cdot 360°}{n}\right)$$

where $k = 0, 1, 2, \ldots, n - 1$.

PROOF   Assuming De Moivre's theorem is true when n is the reciprocal of a positive integer, we have

$$\sqrt[n]{r(\cos \theta + i \sin \theta)} = [r(\cos \theta + i \sin \theta)]^{1/n}$$
$$= r^{1/n}\left(\cos \frac{\theta}{n} + i \sin \frac{\theta}{n}\right)$$

Since $\cos \theta$ and $\sin \theta$ are periodic functions (Section 37) with a period of $360°$, we can say that $\cos \theta = \cos (\theta + k \cdot 360°)$ and $\sin \theta = \sin (\theta + k \cdot 360°)$, where k is an integer. Hence

$$\sqrt[n]{r(\cos \theta + i \sin \theta)}$$
$$= \sqrt[n]{r}\left(\cos \frac{\theta + k \cdot 360°}{n} + i \sin \frac{\theta + k \cdot 360°}{n}\right)$$

It is easy to show that the right side of this equation takes on n distinct values when k takes on the values $0, 1, 2, \ldots, n - 1$. But if k takes on a value larger than

$(n - 1)$, the result is merely a duplication of one of the n roots already found.

EXAMPLE   Find the three cube roots of $-8i$.

SOLUTION   After plotting the number and putting it in trigonometric form, we have

$$-8i = 8(\cos 270° + i \sin 270°)$$

Apply the theorem on roots. The three cube roots of $-8i$ are

$$\sqrt[3]{8}\left(\cos \frac{270° + k \cdot 360°}{3} + i \sin \frac{270° + k \cdot 360°}{3}\right)$$

$$= 2[\cos (90° + k \cdot 120°) + i \sin (90° + k \cdot 120°)]$$

Let the three roots be r_1, r_2, r_3. Then

$$r_1 = 2(\cos 90° + i \sin 90°) = 2i \qquad (k = 0)$$
$$r_2 = 2(\cos 210° + i \sin 210°) = -\sqrt{3} - i \qquad (k = 1)$$
$$r_3 = 2(\cos 330° + i \sin 330°) = \sqrt{3} - i \qquad (k = 2)$$

The three roots are equally spaced on a circle with radius 2 and center at the origin (Figure 95). Notice that for $k = 3$, we obtain r_1 again.

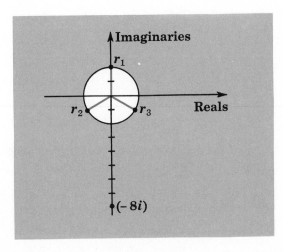

FIGURE 95

EXERCISE 41

Perform the indicated multiplications and then express the results in algebraic form.

1. $4(\cos 110° + i \sin 110°) \cdot 3(\cos 220° + i \sin 220°)$

2. $7(\cos 250° + i \sin 250°) \cdot 5(\cos 290° + i \sin 290°)$

3. $(\cos 50° + i \sin 50°) \cdot 6(\cos 70° + i \sin 70°)$

4. $8(\cos 140° + i \sin 140°) \cdot 2(\cos 190° + i \sin 190°)$

For each of the following products, (a) express the factors in trigonometric form, (b) find their product trigonometrically, and (c) check your result by finding the product algebracially:

5. $7i(4 + 4i)$ **6.** $(-1 - i\sqrt{3})(-\sqrt{3} + i)$

7. $(\sqrt{2} - i\sqrt{2})(-1 + i)$ **8.** $(1 + i\sqrt{3})(-\sqrt{3} - i)$

For each of the following products, (a) express the factors in algebraic form, (b) find their product algebraically, and (c) check your result by finding the product trigonometrically:

9. $6(\cos 150° + i \sin 150°) \cdot 4(\cos 240° + i \sin 240°)$

10. $2(\cos 120° + i \sin 120°) \cdot (\cos 300° + i \sin 300°)$

11. $3(\cos 300° + i \sin 300°) \cdot 8(\cos 330° + i \sin 330°)$

12. $5(\cos 225° + i \sin 225°) \cdot 7(\cos 135° + i \sin 135°)$

Use De Moivre's theorem to find the value of each of the following. Express results in algebraic form.

13. $[\sqrt{10}(\cos 70° + i \sin 70°)]^6$ **14.** $[2(\cos 42° + i \sin 42°)]^5$

15. $(1 + i)^{10}$ **16.** $(2 - 2i)^5$

17. $(-1 - i)^9$ **18.** $(-\sqrt{2} + i\sqrt{2})^8$

19. $\left(-\dfrac{\sqrt{3}}{2} + \dfrac{1}{2}i\right)^{30}$ **20.** $(3 + i\sqrt{3})^4$

21. $(1 - i\sqrt{3})^7$ **22.** $(3 + i)^4$

Find all the indicated roots of the following complex numbers. Express results in algebraic form. Round off approximate results to three-decimal-place accuracy.

23. The cube roots of $125(\cos 333° + i \sin 333°)$

24. The fourth roots of $81(\cos 200° + i \sin 200°)$

25. The square roots of $-18i$

26. The square roots of $32i$

27. The square roots of $-8 - 8i\sqrt{3}$

28. The cube roots of $8i$

29. The cube roots of -1000

30. The cube roots of $4\sqrt{2} - 4i\sqrt{2}$

31. The fourth roots of 16

32. The fifth roots of $-16\sqrt{3} + 16i$

33. The fourth roots of $-8 + 8i\sqrt{3}$

34. The fourth roots of $-10,000i$

Find all the roots of the following equations:

35. $x^6 + 64 = 0$ *Hint:* The roots of the equation $x^6 + 64 = 0$ are the six sixth roots of -64.

36. $x^4 + 256 = 0$

37. $x^5 - 243 = 0$

38. $x^{10} = 1$

39. Prove that

$$\frac{r_1(\cos\theta_1 + i\sin\theta_1)}{r_2(\cos\theta_2 + i\sin\theta_2)} = \frac{r_1}{r_2}[\cos(\theta_1 - \theta_2) + i\sin(\theta_1 - \theta_2)]$$

APPENDIX

83 CIRCULAR AND EXPONENTIAL FUNCTIONS

In calculus it is shown that the exponential function e^{θ} and the circular functions $\cos \theta$ and $\sin \theta$ can be represented by *convergent infinite series.** These are

$$e^{\theta} = 1 + \theta + \frac{\theta^2}{2!} + \frac{\theta^3}{3!} + \frac{\theta^4}{4!} + \frac{\theta^5}{5!} + \cdots + \frac{\theta^n}{n!} + \cdots$$

$$\cos \theta = 1 - \frac{\theta^2}{2!} + \frac{\theta^4}{4!} - \frac{\theta^6}{6!} + \cdots + (-1)^n \frac{\theta^{2n}}{(2n)!} + \cdots$$

$$\sin \theta = \theta - \frac{\theta^3}{3!} + \frac{\theta^5}{5!} - \frac{\theta^7}{7!} + \cdots + (-1)^n \frac{\theta^{2n+1}}{(2n+1)!} + \cdots$$

It is series such as these† that are used in constructing tables like those appearing at the back of this book. For example, with $\theta = 0.178$, and using only two terms of the cosine and sine series, we have

$$\cos 0.178 \doteq 1 - \frac{(0.178)^2}{2} \doteq 1 - 0.0158 = 0.9842$$

$$\sin 0.178 \doteq 0.178 - \frac{(0.178)^3}{6} \doteq 0.178 - 0.0009 = 0.1771$$

---

*Speaking loosely, as we include more and more terms of a series, their sum more closely approaches the expression on the left side. In the case of these three series, this statement is true for all values of θ.

† Notice that $\cos \theta$, an *even* function, is expressed as a series of even powers of θ. Also, the *odd* function $\sin \theta$ is expressed as a series of odd powers of θ.

which agree, to four decimal places, with the corresponding entries in Table 1. These same series are also used by calculators in approximating trigonometric function values, but many more terms of the series are used.

Regardless of how these series were derived, we can take them as alternative definitions of functions that we shall designate by e^θ, $\cos \theta$, and $\sin \theta$, respectively—a far cry from triangles.

Except for the alternating signs in the series for $\sin \theta$ and $\cos \theta$, the sum of these series is the same as the series for e^θ. This prompts us to substitute $i\theta$ for θ in the series for e^θ, which yields

$$e^{i\theta} = 1 + (i\theta) + \frac{(i\theta)^2}{2!} + \frac{(i\theta)^3}{3!} + \frac{(i\theta)^4}{4!} + \frac{(i\theta)^5}{5!} + \cdots + \frac{(i\theta)^m}{m!} + \cdots$$

$$= 1 + i\theta - \frac{\theta^2}{2!} - i\frac{\theta^3}{3!} + \frac{\theta^4}{4!} + i\frac{\theta^5}{5!} + \cdots$$

$$+ (-1)^n \frac{\theta^{2n}}{(2n)!} + i(-1)^n \frac{\theta^{2n+1}}{(2n+1)!} + \cdots$$

$$= \left[1 - \frac{\theta^2}{2!} + \frac{\theta^4}{4!} - \cdots + (-1)^n \frac{\theta^{2n}}{(2n)!} + \cdots \right]$$

$$+ i \left[\theta - \frac{\theta^3}{3!} + \frac{\theta^5}{5!} - \cdots + (-1)^n \frac{\theta^{2n+1}}{(2n+1)!} + \cdots \right]$$

$$= \cos \theta + i \sin \theta$$

Thus in strictly nonrigorous fashion, but with heuristic intent, we have obtained *Euler's formula:*

$$e^{i\theta} = \cos \theta + i \sin \theta$$

An important special case, $\theta = \pi$, yields the unusual relation

$$e^{i\pi} + 1 = 0$$

which involves 0, the real unit 1, the imaginary unit i, and two famous irrational numbers, e and π. Moreover, since $e^{\pi i} = -1$, it follows that $\log_e (-1) = \pi i$.*

---

*The general expression is $(1 + 2k)\pi i$, where k is an integer. This is in partial explanation of the statement that "negative numbers do not have real logarithms."

Raising both sides of Euler's formula to the power n, we get $(e^{i\theta})^n = (\cos\theta + i\sin\theta)^n$. But $(e^{i\theta})^n = e^{in\theta} = \cos n\theta + i\sin n\theta$. Hence we have an alternative derivation of De Moivre's theorem:

$$(\cos\theta + i\sin\theta)^n = \cos n\theta + i\sin n\theta$$

EXERCISE 42

1. Use the series definition of $\cos\theta$ to prove that $\cos(-\theta) = \cos\theta$.
2. Use the series definition of $\sin\theta$ to prove that $\sin(-\theta) = -\sin\theta$.
3. Use the series for $\sin\theta$ to compute $\sin 0.1$ correct to 8 decimal places.
4. Use Euler's formula to derive the formulas for $\sin(A + B)$ and $\cos(A + B)$. *Hint:* Replace θ with $A + B$; notice that $e^{A+B} = e^A e^B$; apply the definition of equality of complex numbers.
5. Use Euler's formula to prove that

   (a) $\cos\theta = \dfrac{e^{i\theta} + e^{-i\theta}}{2}$

   (b) $\sin\theta = \dfrac{e^{i\theta} - e^{-i\theta}}{2i}$

6. Use formula (a) of Problem 5 to prove that

$$\cos 5\theta = 16\cos^5\theta - 20\cos^3\theta + 5\cos\theta$$

Hint: Express the right side in terms of exponential functions; use the binomial formula to expand; simplify to $\frac{1}{2}(e^{5i\theta} + e^{-5i\theta})$, which is $\cos 5\theta$.

84 A GEOMETRIC PROOF, FOR ACUTE ANGLES, OF THE FORMULAS FOR $\sin(A + B)$ AND $\cos(A + B)$

Let A and B be any two positive acute angles whose sum, $A + B$, is also acute. Place A in standard position and place B so that its initial side coincides with the terminal side of A and its vertex falls at the origin O (Figure 96). Choose P as any point on the terminal side of angle $(A + B)$, which is in standard position. From P drop a perpendicular to the initial

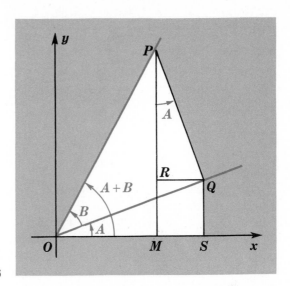

FIGURE 96

side of B at Q. Draw PM and QS perpendicular to the x axis, and draw QR perpendicular to PM. Then angle RPQ equals angle A because they are acute angles with their sides respectively perpendicular. Using the definition of the sine of an angle, we have

$$\sin (A + B) = \frac{MP}{OP} = \frac{MR + RP}{OP} = \frac{SQ + RP}{OP}$$

$$= \frac{SQ}{OP} + \frac{RP}{OP}$$

$$= \frac{SQ}{*} \cdot \frac{*}{OP} + \frac{RP}{\dagger} \cdot \frac{\dagger}{OP}$$

$$= \frac{SQ}{OQ} \cdot \frac{OQ}{OP} + \frac{RP}{QP} \cdot \frac{QP}{OP}$$

$$= \sin A \cos B + \cos A \sin B$$

Using the definition of the cosine of an angle, we have

---

* Since SQ and OP lie in different triangles SOQ and POQ, we multiply top and bottom of the fraction by OQ, the common side of the two triangles.

† The common side of triangles RQP and OQP is QP.

$$\cos (A + B) = \frac{OM}{OP} = \frac{OS - MS}{OP} = \frac{OS - RQ}{OP}$$

$$= \frac{OS}{OP} - \frac{RQ}{OP}$$

$$= \frac{OS}{OQ} \cdot \frac{OQ}{OP} - \frac{RQ}{QP} \cdot \frac{QP}{OP}$$

$$= \cos A \cos B - \sin A \sin B$$

These proofs are not general because we considered only the case in which A and B are positive acute angles with a sum of less than $90°$. A general proof is given in Section 43.

85 COMMON LOGARITHMS

As stated in Chapter 10, the *common,* or *Briggs,* system of logarithms, using the base 10, is most convenient for computation. Therefore, in this text, when the base of a logarithm is not specified, we are to understand that the base is 10. Thus $\log N$ means $\log_{10} N$. And, unless stated to the contrary, the word *logarithm* shall mean common logarithm.

Logarithms are used to shorten the labor involved in computing products and quotients, raising to a power, and extracting roots. For example, the operation of multiplying 543.2 by 6789 can be reduced to the operation of adding the logarithms of these numbers, namely, 2.7350 and 3.8318.

86 CHARACTERISTIC AND MANTISSA

We know that

$\log 1000$	$= 3$	because	10^3	$= 1000$
$\log 100$	$= 2$	because	10^2	$= 100$
$\log 10$	$= 1$	because	10^1	$= 10$
$\log 1$	$= 0$	because	10^0	$= 1$
$\log 0.1$	$= -1$	because	10^{-1}	$= 0.1$
$\log 0.01$	$= -2$	because	10^{-2}	$= 0.01$
$\log 0.001$	$= -3$	because	10^{-3}	$= 0.001$

It seems reasonable to assume that, as a number increases, its logarithm increases.* Consequently, any number lying between 10 and 100 must

---

*This is proved in more advanced texts.

have a logarithm between 1 and 2. This logarithm can be written, then, in the form 1 plus a positive decimal.* For example, $\log 45.7 = 1 + .6599 = 1.6599$. Also, any number between 0.001 and 0.01 must have a logarithm between -3 and -2. This logarithm can be written in the form -3 plus a positive decimal. For example, $\log 0.006 = -3 + .7782$, which can be written $\log 0.006 = 7.7782 - 10$ because $-3 = 7 - 10$.

As a matter of convenience, *every logarithm is usually written as the sum of an integer* (positive, negative, or zero) *plus a positive decimal.* * The integer is called the **characteristic** of the logarithm; the positive decimal is called the **mantissa** of the logarithm. This is illustrated in the following table.

Logarithm	Characteristic	Mantissa
4.5678	4	.5678
0.2345	0	.2345
3.0000	3	.0000
8.7766 − 10 or −2 + .7766	8 − 10 or −2	.7766
7.1111 − 10 or −2.8889	7 − 10 or −3	.1111

87 METHOD OF DETERMINING CHARACTERISTICS

If a number has a decimal point immediately to the right of its first nonzero digit, then the decimal point is said to be in *standard position.*

For example, the decimal point is in standard position in each of the following: 6.507, 4.17, 3.2, and 8. Consequently, *if a number N has its decimal point in standard position,* then N is between 1 and 10, and $\log N$ is between 0 and 1; therefore, *the characteristic of $\log N$ is 0.*

THEOREM 1   Whenever a number is multiplied by 10, its logarithm is increased by 1.

PROOF   Let $\log N$ be the logarithm of any number N. Then

$$\log 10N = \log 10 + \log N \qquad \text{(Property 1)}$$
$$\log 10N = 1 \ + \log N$$

---

*Positive decimal here means a number n such that $0 \leq n < 1$.

It is therefore apparent that when a number is multiplied by 10 (i.e., if the decimal point is moved one place to the right), the characteristic of its logarithm is increased by 1, but the mantissa is unaltered.

ILLUSTRATION   If log 2.345 = 0.3701, then log 23.45 = 1.3701.

By repeating this process, we see that if a number is *multiplied* by 10^k (i.e., if the decimal point is moved k places to the right), the characteristic of its logarithm is *increased* by k. It also follows that if a number is *divided* by 10^k (i.e., if the decimal point is moved k places to the left), the characteristic of its logarithm is *decreased* by k.

ILLUSTRATION   If log 2.345 = 0.3701, then

$$\log 234.5 = 2.3701$$
$$\log 23450 = 4.3701$$
$$\log 0.2345 = 9.3701 - 10$$
$$\log 0.002345 = 7.3701 - 10$$

We may sum up our discussion in the following:

THEOREM 2   If the decimal point in a number is k places to the $\begin{Bmatrix} \text{right} \\ \text{left} \end{Bmatrix}$ of standard position, then the characteristic of the logarithm of the number is $\begin{Bmatrix} k \\ -k \end{Bmatrix}$.

ILLUSTRATION   The characteristic of log 8765 is 3 because the decimal point is understood to be after the 5, which is 3 places to the right of standard position (i.e., after the 8).

The characteristic of log 0.08765 is -2 or $8 - 10$ because the decimal point is 2 places to the left of standard position.

Number	Characteristic of logarithm
45,670	4
456.7	2
4.567	0
0.4567	9 − 10 or −1
0.04567	8 − 10 or −2
0.0004567	6 − 10 or −4

An alternative method used in finding characteristics is

1. For a number N that is greater than 1, the characteristic is one less than the number of digits to the left of the decimal point in N.
2. For a number N that is less than 1, the characteristic is negative and is numerically equal to one more than the number of zeros between the decimal point and the first nonzero digit in N.

THEOREM 3   The mantissa of the logarithm of a number N is independent of the position of the decimal point in N.

This means that two numbers differing only in the position of the decimal point have logarithms with the same mantissa. The proof of this theorem is embodied in that of Theorem 1 and the discussion that follows it.

ILLUSTRATION   The following numbers have logarithms with the same mantissa:

$$0.04689 \qquad 4.689 \qquad 46.89 \qquad 46890$$

Theorem 2 serves two purposes:

1. If we look at the position of the decimal point in a *number,* we can determine the characteristic of its *logarithm.*
2. If we look at the characteristic of the *logarithm* of a number, we can determine the position of the decimal point in the *number.*

EXAMPLE 1   Given $\log 1.616 = 0.2084$. Find:

  (a) $\log 0.01616$
  (b) N if $\log N = 5.2084$
  (c) N if $\log N = 6.2084 - 10$

SOLUTION   (a) $\log 0.01616 = 8.2084 - 10$

Since the decimal point in 0.01616 is **2** places to the *left* of standard position, the characteristic is -2 or $8 - 10$. The mantissa is the same as that for $\log 1.616$.

  (b) If $\log N = 5.2084$
            $N = 161{,}600$

Since $\log N$ and $\log 1.616$ have the same mantissa, N is obtainable from 1.616 by moving the decimal point 5 places to the right (from standard position).

(c) If
$$\log N = 6.2084 - 10$$
$$N = 0.000\ 1616$$

Since the characteristic of $\log N$ is -4, we obtain N from 1.616 by moving the decimal point 4 places to the left.

Note Theorems 1, 2, and 3 of this section are valid only if the base of the logarithms is 10. For any other base, the process of finding the characteristic would not be so simple.

88 A FOUR-PLACE TABLE OF MANTISSAS

In Table 2 there are listed, to four decimal places, the mantissas (with the decimal points omitted) of the logarithms of all positive integers from 1 to 999. Since the mantissa is independent of the decimal point in the number, Table 2 can be used to find the mantissa of the logarithm of any three-figure number. The problems we shall need to consider are

1. Given a number N, to find $\log N$.
2. Given $\log N$, to find N.

The procedure of finding the logarithm of a given number is illustrated by the following example.

EXAMPLE 1 Find $\log 0.0526$.

SOLUTION The characteristic of $\log 0.0526$ is -2 or $8 - 10$. To find the mantissa, look for 52 in the left-hand column headed N in Table 2. In the line beginning with 52, move over to the column headed by 6 and find 7210. Hence

$$\log 0.0526 = 8.7210 - 10$$

The procedure of finding the number whose logarithm is given is illustrated by the following examples.

EXAMPLE 2 Given log $N = 1.3345$, find N.

SOLUTION Look for the mantissa .3345 in the body of Table 2. It appears in the 21 line and the 6 column. Hence N is 216, with the decimal point placed in accordance with a characteristic of 1. Therefore,

if $$\log N = 1.3345$$

then $$N = 21.6$$

EXAMPLE 3 Given log $N = 7.0969 - 10$, find N.

SOLUTION The mantissa .0969 appears in the 12 row and the 5 column. Hence N is 125 with the decimal point moved 3 places to the left of standard position (because the characteristic is $7 - 10 = -3$). Therefore,

if $$\log N = 7.0969 - 10$$

then $$N = 0.00125$$

EXERCISE 43

Use a four-place table to find the value of each of the following:

1. log 82.1
2. log 0.000 0367
3. log 738,000
4. log 0.351
5. log 0.0189
6. log 9.15
7. log 0.000 271
8. log 9,490,000
9. log 6460
10. log 59,100
11. log 4.53
12. log 0.00297

Use a four-place table to find the value of N to three significant digits.

13. $\log N = 5.8525 - 10$
14. $\log N = 3.4624$
15. $\log N = 2.5263$
16. $\log N = 0.8082$
17. $\log N = 6.6721$
18. $\log N = 9.9186 - 10$
19. $\log N = 7.8457 - 10$
20. $\log N = 8.6628 - 10$

21. $\log N = 0.9800$

22. $\log N = 4.7536 - 10$

23. $\log N = 4.2553$

24. $\log N = 1.7143$

89 INTERPOLATION

We have seen that the logarithm of a three-figure number can be found by use of Table 2. If a number is increased by a small amount, the change in its logarithm is very nearly proportional to the change in the number itself. Hence the logarithm of a four-figure number can be found from Table 2 by interpolation. It is to be remembered that, since the mantissas of the logarithms of most integers are unending decimals, a four-place table merely rounds off these mantissas to four-figure accuracy. Hence the results obtained in using a four-place table will usually be only approximations, with the last figure subject to error.

EXAMPLE 1 Find log 182.7.

SOLUTION The characteristic is 2. Since the mantissa of the logarithm is independent of the position of the decimal point in the number, we see that the required mantissa lies between the mantissas for 1820 and 1830.

$$\begin{array}{lll} \text{Mantissa of log* 1820} & = .2601 \\ \text{Mantissa of log } 1827 & = \\ \text{Mantissa of log } 1830 & = .2625 \end{array} \quad 7 \quad 10 \quad 24$$

As the number increases by 10 (from 1820 to 1830), the mantissa increases 24 ten-thousandths. Our number is $\frac{7}{10}$ of the way from 1820 to 1830. Hence the required mantissa is $\frac{7}{10}$ of the way from .2601 to .2625. But $\frac{7}{10}(24) = 16.8 \rightarrow 17$. Add the 17 ten-thousandths to .2601 to get .2618, the required mantissa. Hence

$$\log 182.7 = 2.2618$$

EXAMPLE 2 Given $\log N = 9.2950 - 10$, find N.

---

*A convenient abbreviation for *mantissa of log* is *m-log*.

SOLUTION We search for the mantissa .2950 in the body of Table 2. It lies between the mantissas .2945 and .2967.

$$
\begin{array}{lll}
m\text{-log } 1970 & = .2945 \\
m\text{-log } N & = .2950 \\
m\text{-log } 1980 & = .2967
\end{array}
\quad 10 \quad \quad 5 \quad 22
$$

The given mantissa is $\frac{5}{22}$ of the way from .2945 to .2967. Hence the required number (aside from decimal point) should be $\frac{5}{22}$ of the way from 1970 to 1980. But $\frac{5}{22}(10) = 2\frac{3}{11} \to 2$. Adding 2 to 1970, we find N is 1972 with the decimal point placed in accordance with a characteristic of $9 - 10$. Hence

if $\qquad\qquad \log N = 9.2950 - 10$

then $\qquad\qquad N = 0.1972$

EXERCISE 44

Use a four-place table to find the value of each of the following:

1. log 0.003032
2. log 47.14
3. log 0.000 05498
4. log 7015
5. log 967.2
6. log 0.02493
7. log 82.57
8. log 0.000 1249
9. log 0.2923
10. log 6,385,000
11. log 0.09869
12. log 884.1
13. log 57,160
14. log 0.000 7504
15. log 3.116
16. log 0.000 04638

Use a four-place table to find the value of N to four significant digits.

17. $\log N = 4.9347 - 10$
18. $\log N = 2.6061$
19. $\log N = 3.9659$
20. $\log N = 9.8641 - 10$
21. $\log N = 5.4758$
22. $\log N = 7.8068 - 10$
23. $\log N = 9.7410 - 10$
24. $\log N = 5.5743$
25. $\log N = 6.2736 - 10$
26. $\log N = 0.1311$
27. $\log N = 6.0087$
28. $\log N = 4.4615 - 10$

29. $\log N = 1.5399$ **30.** $\log N = 3.7097$

31. $\log N = 5.8313 - 10$ **32.** $\log N = 4.9497$

90 LOGARITHMIC COMPUTATION

When logarithms are used to compute products, quotients, and powers of numbers, it is advisable to

1. Make a complete outline indicating the operations to be performed.
2. Fill in all characteristics.
3. Fill in all mantissas.
4. Perform the operations outlined in step 1.

These suggestions are offered in the hope that accuracy, speed, and neatness will be achieved. Every logarithm appearing in the solution should be labeled.

EXAMPLE 1 Use logarithms to compute $\dfrac{(2460)(0.357)}{8.18}$.

SOLUTION Let
$$N = \frac{(2460)(0.357)}{8.18}$$

Then $\log N = \log 2460 + \log 0.357 - \log 8.18$. After preparing the outline and filling in the characteristics, we have

$$
\begin{aligned}
\log 2460 &= 3. \\
\log 0.357 &= 9. \qquad\quad - 10 \\
\log \text{numerator} &= \underline{\hspace{3cm}} \quad \text{Add} \\
\log 8.18 &= 0. \\
\log N &= \underline{\hspace{3cm}} \quad \text{Subtract} \\
N &=
\end{aligned}
$$

After supplying the mantissas and performing the indicated operations, we have

$$
\begin{aligned}
\log 2460 &= 3.3909 \\
\log 0.357 &= \underline{9.5527 - 10} \\
\log \text{numerator} &= 12.9436 - 10 \quad \text{Add} \\
\log 8.18 &= \underline{0.9128} \\
\log N &= 2.0308 \quad\quad \text{Subtract} \\
N &= 107
\end{aligned}
$$

Notice that no interpolation was performed in finding N from log N. The original numbers are all three-figure numbers; hence the computed result should have no more than three-figure accuracy. Since the mantissa .0308 is best approximated by the tabular mantissa .0294, the best three-figure approximation of N is 107.

In all logarithmic computations we are really expressing the original numbers as powers of 10. For example, since log 2460 = 3.3909, it follows that 2460 = $10^{3.3909}$. Consequently

$$N = \frac{(10^{3.3909})(10^{9.5527-10})}{10^{0.9128}}$$

Apply the laws of exponents:

$$N = 10^{3.3909+(9.5527-10)-0.9128} = 10^{2.0308} = 107$$

EXAMPLE 2 Use logarithms to compute

$$N = \frac{(1.789)^3}{(87,650)(0.04466)}$$

SOLUTION   Take the logarithm of each side:

$$\log N = 3 \log 1.789 - (\log 87,650 + \log 0.04466)$$

After taking the four suggested steps, we get

log 1.789 = $\underline{0.2526}$ 3	log 87,650 = 4.9428
3 log 1.789 = 0.7578	log 0.04466 = $\underline{8.6499 - 10}$ A
log num. = 10.7578 − 10	log den. = 13.5927 − 10
log den. = $\underline{3.5927}$ ← S	= 3.5927
log N = 7.1651 − 10	
N = 0.001462	

Notice that log numerator was changed from 0.7578 to 10.7578 − 10 to avoid subtracting 3.5927 from a smaller number.

This example illustrates four-figure accuracy in the original data and in the computed result.

EXAMPLE 3 Compute $\sqrt[3]{\dfrac{0.00559}{90.16}}$.

SOLUTION   Let $N = \sqrt[3]{\dfrac{0.00559}{90.16}}$.

Then $\log N = \tfrac{1}{3}(\log 0.00559 - \log 90.16)$

After taking the four suggested steps, we have

$$
\begin{aligned}
\log 0.00559 &= 7.7474 - 10 \\
\log 90.16 &= 1.9550 \\
\log \text{radicand} &= 5.7924 - 10 \text{ S}\\
\log \text{radicand} &= 25.7924 - 30 \\
3 &\overline{} \\
\log N &= 8.5975 - 10 \\
N &= 0.0396
\end{aligned}
$$

Notice that log radicand was changed from $5.7924 - 10$ to $25.7924 - 30$ to facilitate the division by 3. Had we divided $5.7924 - 10$ by 3, the result, $1.9308 - 3.3333 = -1.4025$, would involve a *negative* decimal that does not occur in our table of *positive* mantissas.

The final result is written with only three-figure accuracy because the least accurate number, 0.00559, in the original data has only three significant figures.

EXAMPLE 4 Use logarithms to compute

$$
x = \frac{(-1.789)^3}{(-87,650)(-0.04466)}
$$

SOLUTION   The value of x is negative, since $\dfrac{(-)^3}{(-)(-)} = \dfrac{-}{+} = -$. Discard all minus signs in x, and then use logarithms to compute the value of the corresponding expression in which all numbers are positive. This was done in Example 2 with the result 0.001462. Hence

$$
x = -0.001462
$$

EXERCISE 45

Use logarithms to compute the following, correct to three-figure accuracy. (In finding N from log N, do not interpolate.)

1. $(678)(0.192)$ **2.** $(4350)(0.0629)(0.00175)$

3. $(2.50)^{10}$ **4.** $(0.721)^6(83.4)$

5. $\dfrac{(4.05)^3}{136,000}$

6. $\left(\dfrac{707}{99.3}\right)^2$

7. $[(83.9)(0.00765)]^7$

8. $\dfrac{5.29}{0.000\ 444}$

9. $\sqrt[4]{(81.2)(3740)}$

10. $(0.576)\sqrt[8]{30,000}$

11. $\dfrac{\sqrt{0.0471}}{0.000\ 958}$

12. $\sqrt[5]{\dfrac{0.00666}{7310}}$

13. $\sqrt[9]{\dfrac{0.00259}{0.0406}}$

14. $\sqrt[3]{0.000\ 828}$

15. $\sqrt[6]{(203)(0.00174)}$

16. $\sqrt[7]{0.00285}$

17. $(-35,790)(-0.0202)^4$

18. $[(-1.42)(-208)(-0.000\ 641)]^5$

19. $\dfrac{-56,700}{(-64.3)(-819)}$

20. $\left[\dfrac{(-95,900)(-0.0107)}{-328}\right]^7$

21. $\sqrt{\log 8710}$

22. $\dfrac{\log 537}{\log 9.12}$

23. $(\log 0.302)^9$

24. $(\log 2.63)(\log 706)$

25. $(3.49)^{-3}$

26. $\dfrac{0.841}{(4.06)^4 - 171.6}$

27. $\dfrac{\sqrt[3]{8490} + 9.60}{751}$

28. $(23,400)^{2/3}$

Use logarithms to compute the following, correct to four-figure accuracy:

29. $\sqrt[100]{26,300}$

30. $(0.6901)^{-20}$

31. $(125,400)^{7/10}$

32. $\dfrac{0.000\ 08004}{(0.3199)^3}$

33. $\dfrac{(0.008506)^2}{(0.1479)^9}$

34. $\sqrt[4]{(865,400)(9003)}$

35. $\dfrac{(76,210)\sqrt[8]{5030}}{4.890}$

36. $\sqrt[10]{0.0005160}$

Use logarithms to compute the following to as much accuracy as is warranted by the numbers involved:

37. $(58.3)(712)(0.000\ 0209)$

38. $\dfrac{(22,300)(0.004444)}{(77.80)(0.05981)}$

39. $(1.08)^{30}$

40. $\left(\dfrac{1850}{972}\right)^{-4}$

41. $\dfrac{746.0}{\sqrt[3]{0.000\ 9580}}$

42. $\dfrac{\sqrt[5]{0.00283}}{\sqrt{76.50}}$

43. $\dfrac{71.9}{(5260)(0.43)}$

44. $\sqrt{648,000\ \sqrt[3]{0.0057}}$

45. The time t in seconds for one complete oscillation of a pendulum of length l centimeters is given by $t = 2\pi\sqrt{\dfrac{l}{g}}$, where $g = 980.7$ centimeters per second per second. Find t for a pendulum 102.5 cm long.

46. For a certain gas at a certain temperature, the pressure p (pounds per square inch) and the volume v (cubic inches) satisfy the equation $pv^{1.4} = 25,000$. Find the pressure on the gas if its volume is 730 in^3.

47. At 3-month intervals, a person deposits $1000 in a savings and loan association that pays 8 percent interest, compounded quarterly. A mathematical formula for the amount S of this annuity at the end of 20 years (80 quarters) gives

$$S = 1000\left[\frac{(1.02)^{80} - 1}{0.02}\right]$$

Compute S to four-figure accuracy.

48. If a person deposits a certain sum of money A at a bank that pays 6 percent interest, compounded semiannually, the bank will then pay the customer $5000 every 6 months for a period of 10 years. Using

$$A = 5000\left[\frac{1 - (1.03)^{-20}}{0.03}\right]$$

compute A to four-figure accuracy.

91 EXPONENTIAL EQUATIONS

An exponential equation is an equation in which a variable appears in an exponent. Such an equation can usually be solved by equating the logarithms of the two sides and then finding the roots of the resulting algebraic equation.

EXAMPLE   Solve for x: $(9.55)^x = 0.0345$

SOLUTION   Take the logarithm of each side:

$$\log (9.55)^x = \log 0.0345$$

$$x \log 9.55 = \log 0.0345$$

$$x = \frac{\log 0.0345}{\log 9.55}$$

$$= \frac{8.5378 - 10}{0.9800}$$

$$= \frac{-1.4622}{0.98}$$

$$= -1.49$$

In this case it is easier to perform the division without using logarithms. Had logarithms been used, we should first have computed the value of the fraction $\dfrac{1.4622}{0.98}$ and then attached a minus sign to the result.

It should be noted that $\dfrac{\log 0.0345}{\log 9.55}$ is the quotient of two logarithms, not the logarithm of a quotient.

92 CHANGE OF BASE OF LOGARITHMS

In making numerical computations, the most convenient system of logarithms is the *common*, or *Briggs*, system, which employs the base 10. If we know the logarithm of a number to the base a, we can find the logarithm of that number to the base b by using

$$\log_b N = \frac{\log_a N}{\log_a b} = (\log_b a)(\log_a N) \tag{1}$$

To prove this,

let $$\log_b N = y$$

Then $$N = b^y$$

Take the logarithm of each side to the base a:

$$\log_a N = \log_a b^y$$
$$= y \log_a b$$
$$\log_a N = \log_b N \log_a b$$

Hence
$$\log_b N = \frac{\log_a N}{\log_a b} \qquad (2)$$

If $N = a$,

$$\log_b a = \frac{1}{\log_a b}$$

Therefore

$$\log_b N = (\log_b a)(\log_a N)$$

In calculus and higher mathematics, the most suitable system of logarithms is the *natural,* or *napierian,* system, which employs the base e, where e is approximately 2.71828. If $a = 10$ and $b = e$, Equation (1) becomes

$$\log_e N = \frac{\log_{10} N}{0.43429} = 2.3026 \log_{10} N \qquad (3)$$

The usual abbreviation for $\log_e N$ is $\ln N$.

Hence
$$\ln N = 2.3026 \log_{10} N$$

Thus the natural logarithm of a number may be obtained by multiplying its common logarithm by 2.3026.

EXERCISE 46

Use Table 2 to solve for x.

1. $(5.37)^x = 0.482$
2. $(12.6)^x = 929$
3. $(0.871)^x = 0.0604$
4. $(0.00302)^x = 0.855$

5. $(0.0263)^x = 13.9$ **6.** $(0.708)^x = 0.00597$

7. $(ab)^x = c^{x+1}$ **8.** $a^{3x-4} = b^{5x+6}$

9. $y = \dfrac{10^x + 10^{-x}}{2}$ *Hint:* Multiply both sides by $2(10^x)$ to get a quad-

ratic equation in 10^x.

10. $(30.9)^x(9.12)^{x+1} = 1000$

11. $(5.61)^x = (413)(0.798)^{3x+1}$

12. Use substitution to check each solution of the equation in Problem 9.

Evaluate each of the following:

13. $\log_e 100$ **14.** $\log_e 0.742$ **15.** $\log_e 0.0583$

16. $\log_e 8.66$ **17.** $\log_5 0.041$ **18.** $\log_3 10.0$

19. $\log_2 520$ **20.** $\log_4 0.175$

21. The half-life of radium is about 1700 years. (That is, after 1700 years, only half of a present supply of radium will remain; the rest will have disintegrated.) If the initial amount of radium is 100 milligrams and N is the number of milligrams that are left after t years, then

$$N = 100(\tfrac{1}{2})^{t/1700}$$

(a) How much radium will be present after 200 years?

(b) How many years must pass before only 90 milligrams remain?

22. When bacteria grow under ideal conditions, their rate of growth varies directly as the number of bacteria present. If the initial number of bacteria in a certain culture is 1000, and N is the number after t hours, then

$$N = 1000(2^t)$$

(a) How many bacteria will be present after 1.60 hours?

(b) How many hours must pass before 9000 bacteria are present?

93 LOGARITHMS OF TRIGONOMETRIC FUNCTIONS

Table 3 lists, to four decimal places, the logarithms of the sine, cosine, tangent, and cotangent for acute angles at intervals of $\frac{1}{10}$ of a degree. This table is a combination of Tables 1 and 2. We could find the value of log sin 33.3° by using Table 1 to find sin 33.3° = 0.5490, then using Table 2 to find log 0.5490 = 9.7396 − 10. It is much easier, however, to use Table 3 and read immediately

$$\log \sin 33.3° = 9.7396 - 10$$

The sine or cosine of any acute angle is less than 1. The same is true of the tangent of an angle between 0° and 45°, or the cotangent of an angle between 45° and 90°. Consequently, the characteristics of the logarithms of such functions are negative. To conserve space in the tables, the "−10" of negative characteristics has been omitted. For the sake of uniformity, the table is so constructed that a "−10" is to be understood with each entry. For example, the table entry for log tan 86.8° is 11.2525. But we are to understand that

$$\log \tan 86.8° = 11.2525 - 10 = 1.2525$$

Table 3 can be used to read directly the logarithms of the trigonometric functions of angles measured to the nearest tenth of a degree. By using interpolation, we can extend this to angles measured to the nearest hundredth of a degree.

EXAMPLE 1 Find log tan 72.58°.

SOLUTION Using Table 3, we get

$$
\begin{array}{ll}
\log \tan 72.50° & = 0.5013 \\
\log \tan 72.58° \quad 8 \quad 10 = & 26 \\
\log \tan 72.60° & = 0.5039
\end{array}
$$

Since

$$\tfrac{8}{10}(26) = 20.8 \rightarrow 21$$
$$\log \tan 72.58° = 0.5034$$

EXAMPLE 2 Find θ if log cos θ = 9.8715 − 10.

SOLUTION From Table 3, we obtain

$$
\begin{array}{ll}
\log \cos 41.90° & = 9.8718 - 10 \\
\log \cos \quad \theta \quad \Big)10 = 9.8715 - 10 \quad 3 \Big) 7 \\
\log \cos 42.00° & = 9.8711 - 10
\end{array}
$$

Inasmuch as $\qquad \frac{3}{7}(10) = 4\frac{2}{7} \rightarrow 4$

if $\qquad\qquad \log \cos \theta = 9.8715 - 10$

then $\qquad\qquad\quad \theta = 41.94°$

EXERCISE 47

Use Table 3 to evaluate the following:

1. log sin 80.1°
2. log cos 2.6°
3. log tan 47.2°
4. log cot 79.1°
5. log cot 32.84°
6. log tan 63.09°
7. log cos 66.12°
8. log sin 42.51°
9. log tan 8.07°
10. log cot 53.83°
11. log sin 72.33°
12. log cos 22.86°
13. log cos 54.28°
14. log sin 11.64°
15. log cot 16.97°
16. log tan 83.82°

Use Table 3 to find the value of θ.

17. $\log \cos \theta = 9.9414 - 10$
18. $\log \sin \theta = 9.8751 - 10$
19. $\log \cot \theta = 9.8708 - 10$
20. $\log \tan \theta = 9.7015 - 10$
21. $\log \tan \theta = 0.0285$
22. $\log \cot \theta = 0.2057$
23. $\log \sin \theta = 9.4763 - 10$
24. $\log \cos \theta = 9.4811 - 10$
25. $\log \cot \theta = 9.1576 - 10$
26. $\log \tan \theta = 8.9780 - 10$
27. $\log \cos \theta = 9.9130 - 10$
28. $\log \sin \theta = 9.9377 - 10$
29. $\log \sin \theta = 9.2635 - 10$
30. $\log \cos \theta = 9.6544 - 10$
31. $\log \tan \theta = 0.5800$
32. $\log \cot \theta = 0.0871$

33. If θ is any angle in Q I, prove that

$$\log \tan \theta + \log \cot \theta = 0$$

94 SOLVING OBLIQUE TRIANGLES: *SAS* AND *SSS*

If we want a high degree of accuracy in the computed parts of an oblique triangle and we have to use logarithms in making the calculations, then it is usually desirable to avoid using the law of cosines in solving the *SAS* and *SSS* cases. The following outline should assist the student in choosing a suitable plan of attack.

	Without logs	With logs
SAA	Law of sines	Law of sines
SSA	Law of sines	Law of sines
SAS	Law of cosines	Law of cosines (or Law of tangents)
SSS	Law of cosines	Law of cosines (or Half-angle formulas)

95 THE LAW OF TANGENTS

In any triangle, the difference of two sides is to their sum as the tangent of half the difference of the opposite angles is to the tangent of half their sum:

$$\frac{a-b}{a+b} = \frac{\tan\frac{1}{2}(A-B)}{\tan\frac{1}{2}(A+B)} \qquad \frac{b-a}{b+a} = \frac{\tan\frac{1}{2}(B-A)}{\tan\frac{1}{2}(B+A)}$$

$$\frac{b-c}{b+c} = \frac{\tan\frac{1}{2}(B-C)}{\tan\frac{1}{2}(B+C)} \qquad \frac{c-b}{c+b} = \frac{\tan\frac{1}{2}(C-B)}{\tan\frac{1}{2}(C+B)}$$

$$\frac{c-a}{c+a} = \frac{\tan\frac{1}{2}(C-A)}{\tan\frac{1}{2}(C+A)} \qquad \frac{a-c}{a+c} = \frac{\tan\frac{1}{2}(A-C)}{\tan\frac{1}{2}(A+C)}$$

PROOF Let the two given sides be a and b, with $a > b$. By the law of sines,

$$\frac{a}{b} = \frac{\sin A}{\sin B}$$

Subtract 1 from each side; add 1 to each side:

$$\frac{a}{b} - 1 = \frac{\sin A}{\sin B} - 1 \qquad \frac{a}{b} + 1 = \frac{\sin A}{\sin B} + 1$$

Hence

$$\frac{a-b}{b} = \frac{\sin A - \sin B}{\sin B} \qquad \frac{a+b}{b} = \frac{\sin A + \sin B}{\sin B}$$

Divide the first equation by the second:

$$\frac{a - b}{a + b} = \frac{\sin A - \sin B}{\sin A + \sin B}$$

Apply formulas (16) and (15) of Section 49:

$$\frac{a - b}{a + b} = \frac{2 \cos \frac{1}{2}(A + B) \sin \frac{1}{2}(A - B)}{2 \sin \frac{1}{2}(A + B) \cos \frac{1}{2}(A - B)}$$

$$= \cot \tfrac{1}{2}(A + B) \tan \tfrac{1}{2}(A - B) = \frac{\tan \frac{1}{2}(A - B)}{\tan \frac{1}{2}(A + B)}$$

This proves the first of the six formulas.

If a and b are interchanged, then their opposite angles, A and B, must be interchanged and we get the second formula:

$$\frac{b - a}{b + a} = \frac{\tan \frac{1}{2}(B - A)}{\tan \frac{1}{2}(B + A)}$$

The remaining formulas may be obtained by cyclic permutation (Section 68).

96 APPLICATIONS OF THE LAW OF TANGENTS: *SAS*

Suppose that the given parts are the sides a and b and the included angle C. If $a > b$, we use the following form of the law of tangents:

$$\tan \frac{1}{2}(A - B) = \frac{a - b}{a + b} \tan \frac{1}{2}(A + B)$$

Knowing a and b, we can find $(a - b)$ and $(a + b)$. Since C is given, we can find $\frac{1}{2}(A + B)$ by halving the relation $A + B = 180° - C$. Thus the three quantities on the right side of the equation are easily obtained from the given data. By applying the law of tangents, we find the value of $\frac{1}{2}(A - B)$. *Knowing $\frac{1}{2}(A + B)$ and $\frac{1}{2}(A - B)$, we add to get A, and subtract to get B.* The sixth part, c, can be found with the law of sines. The problem can be checked by finding c again with another form of the law of sines.

EXAMPLE   Solve the triangle ABC, given $a = 567$, $b = 321$, $C = 49.0°$.

SOLUTION   Since $a > b$, use the first form of the law of tangents:

$$\tan \frac{1}{2}(A - B) = \frac{a - b}{a + b} \tan \frac{1}{2}(A + B)$$

$$
\begin{aligned}
a &= 567 \\
b &= 321 \\
\hline
a - b &= 246 \\
a + b &= 888 \\
A + B &= 180° - 49.0° \\
&= 131.0° \\
\tfrac{1}{2}(A + B) &= 65.5° \longrightarrow \tan \tfrac{1}{2}(A - B) = \tfrac{246}{888} \tan 65.5°
\end{aligned}
$$

$$
\begin{aligned}
\log 246 &= 2.3909 \\
\log \tan 65.5° &= 0.3413 \\
\hline
\log \text{num.} &= 12.7322 - 10 \quad \text{A} \\
\log 888 &= 2.9484 \\
\hline
\log \tan \tfrac{1}{2}(A - B) &= 9.7838 - 10 \quad \text{S}
\end{aligned}
$$

$$
\begin{aligned}
\tfrac{1}{2}(A - B) &= 31.3° \longleftarrow \tfrac{1}{2}(A - B) = 31.3° \\
A &= 96.8° \\
B &= 34.2°
\end{aligned}
$$

To find c, $$c = \frac{a \sin C}{\sin A} = \frac{567 \sin 49.0°}{\sin 96.8°}$$

$$
\begin{aligned}
\log 567 &= 2.7536 \\
\log \sin 49.0° &= 9.8778 - 10 \\
\hline
\log \text{num.} &= 12.6314 - 10 \quad \text{A} \\
\log \sin 96.8° &= 9.9969 - 10 \\
\hline
\log c &= 2.6345 \quad \text{S} \\
c &= 431
\end{aligned}
$$

To check, $$c = \frac{b \sin C}{\sin B} = \frac{321 \sin 49.0°}{\sin 34.2°}$$

$$
\begin{aligned}
\log 321 &= 2.5065 \\
\log \sin 49.0° &= 9.8778 - 10 \\
\hline
\log \text{num.} &= 12.3843 - 10 \quad \text{A} \\
\log \sin 34.2° &= 9.7498 - 10 \\
\hline
\log c &= 2.6345 \quad \text{S} \\
c &= 431
\end{aligned}
$$

EXERCISE 48

Make a complete outline of the logarithmic solution and check of the oblique triangle in which the following parts are known:

1. *a, c, B with a > c*

2. *b, c, A with b > c*

3. *a, b, C with b > a*

4. *a, c, B with c > a*

Solve the following triangles. Check as directed by the instructor.

5. $a = 487, b = 253, C = 66.0°$

6. $a = 83.6, b = 59.4, C = 51.8°$

7. $a = 1.13, b = 7.41, C = 102.6°$

8. $a = 942, b = 678, C = 26.2°$

9. $a = 72.8, c = 65.2, B = 19.4°$

10. $b = 275, c = 437, A = 126.0°$

11. $a = 3915, c = 5166, B = 76.00°$

12. $a = 16.93, b = 80.40, C = 35.80°$

13. Two sides of a triangle are 33 and 44. The difference in the angles opposite these sides is 68°. Find the angles.

14. The sum of the two shortest sides of a triangle is 100 cm. Two angles of the triangle are 110° and 50°. Find the two shortest sides of the triangle.

97 THE HALF-ANGLE FORMULAS

In any triangle ABC,

$$\tan \frac{A}{2} = \frac{r}{s-a} \qquad \tan \frac{B}{2} = \frac{r}{s-b} \qquad \tan \frac{C}{2} = \frac{r}{s-c}$$

where

$$r = \sqrt{\frac{(s-a)(s-b)(s-c)}{s}} \qquad \text{and} \qquad s = \frac{1}{2}(a+b+c)$$

PROOF   In Section 71 we proved that the area of a triangle with sides a, b, c is $K = \sqrt{s(s-a)(s-b)(s-c)}$, where $s =$

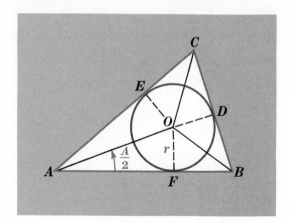

FIGURE 97

$\frac{1}{2}(a + b + c)$. In triangle ABC, let AO, BO, and CO be the bisectors of angles A, B, and C, respectively (Figure 97). Then, by geometry, O is the center of the inscribed circle. *Let r be the radius of the inscribed circle.* It is obvious that

$$\text{area } \triangle ABC = \text{area } \triangle AOB + \text{area } \triangle BOC + \text{area } \triangle COA$$
$$= \tfrac{1}{2}cr + \tfrac{1}{2}ar + \tfrac{1}{2}br$$
$$= r \cdot \tfrac{1}{2}(c + a + b) = rs$$

Equate the two values of the area:

$$rs = \sqrt{s(s - a)(s - b)(s - c)}$$
$$r = \sqrt{\frac{(s - a)(s - b)(s - c)}{s}}$$

In $\triangle ABC$, by geometry, $AF = AE$, $BF = BD$, and $CD = CE$. Why? Since the sum of these six segments equals the perimeter, $2s$,

$$2AF + 2BD + 2CD = 2s$$
$$AF = s - (BD + CD) = s - a$$

Similarly

$$BD = s - b \quad \text{and} \quad CE = s - c$$

Since AO bisects angle A and $OF \perp AF$,

$$\tan \frac{A}{2} = \frac{FO}{AF} = \frac{r}{s - a}$$

Similarly

$$\tan\frac{B}{2} = \frac{r}{s-b} \qquad \text{and} \qquad \tan\frac{C}{2} = \frac{r}{s-c}$$

98 APPLICATIONS OF THE HALF-ANGLE FORMULAS: *SSS*

When the three sides are given, we first compute s, then $(s-a)$, $(s-b)$, and $(s-c)$. Knowing these four quantities, we find r by use of logarithms. Then the half-angle formulas give us $A/2$, $B/2$, and $C/2$. When these values are doubled, we get the three angles of the triangle. The half-angle formulas may be used to check other solutions.

EXAMPLE   Solve the triangle ABC, given $a = 76.5$, $b = 60.0$, $c = 54.3$.

SOLUTION

$$
\begin{aligned}
a &= 76.5 \\
b &= 60.0 \\
c &= 54.3 \quad \text{A} \\
2s &= 190.8 \\
s &= 95.4
\end{aligned}
$$

$$
\begin{aligned}
s - a &= 18.9 \\
s - b &= 35.4 \\
s - c &= 41.1 \quad \text{A} \\
s &= 95.4
\end{aligned}
$$

*Check:**

$$r = \sqrt{\frac{(s-a)(s-b)(s-c)}{s}}$$

$$
\begin{aligned}
\log(s-a) &= 1.2765 \\
\log(s-b) &= 1.5490 \\
\log(s-c) &= 1.6138 \quad \text{A} \\
\log \text{num.} &= 4.4393 \\
\log s &= 1.9795 \quad \text{S} \\
\log \text{radicand} &= 2.4598 \\
2 & \;\overline{} \\
\log r &= 1.2299
\end{aligned}
$$

$$\tan\frac{A}{2} = \frac{r}{s-a}$$

$$
\begin{aligned}
\log r &= 1.2299 \\
\log(s-a) &= 1.2765 \quad \text{S} \\
\log \tan\frac{A}{2} &= 9.9534 - 10
\end{aligned}
$$

$$\frac{A}{2} = 41.93°$$

$$A = 83.86°$$

$$\tan\frac{B}{2} = \frac{r}{s-b}$$

$$
\begin{aligned}
\log r &= 1.2299 \\
\log(s-b) &= 1.5490 \quad \text{S} \\
\log \tan\frac{B}{2} &= 9.6809 - 10
\end{aligned}
$$

$$\frac{B}{2} = 25.62°$$

$$B = 51.24°$$

$*(s-a) + (s-b) + (s-c) = 3s - (a+b+c) = s.$

$$\tan \frac{C}{2} = \frac{r}{s-c}$$

$$\log r = 1.2299$$

$$\log (s - c) = \underline{1.6138}$$

$$\log \tan \frac{C}{2} = 9.6161 - 10$$

$$\frac{C}{2} = 22.45°$$

$$C = 44.90°$$

S

Rounding off these results to the nearest tenth of a degree for three-figure accuracy, we get

$$A = 83.9°$$
$$B = 51.2°$$
$$C = 44.9°$$

Check
$$A + B + C = 180.0°$$

Comment. If the sum of the angles in the triangle differs from 180.0° or 180.00° by 1 or 2 in the last retained digit, the check is satisfactory.

EXERCISE 49

1. Make a complete outline of the logarithmic solution of an oblique triangle in which the three sides are known.

Solve the following triangles:

2. $a = 505$, $b = 606$, $c = 707$
3. $a = 29.1$, $b = 31.4$, $c = 42.3$
4. $a = 8.60$, $b = 4.72$, $c = 5.38$
5. $a = 754$, $b = 289$, $c = 915$
6. $a = 1980$, $b = 3460$, $c = 2870$
7. $a = 649.0$, $b = 812.3$, $c = 361.7$
8. $a = 9238$, $b = 5762$, $c = 4900$
9. $a = 6525$, $b = 7001$, $c = 1396$

10. The perimeter of a triangle is 40. The radius of its inscribed circle is 2.0. If one angle of the triangle is 52°, how long is the side opposite it?

11. The perimeter of a triangle is 12. The radius of its inscribed circle is

1. If one side of the triangle is 3, find the other two sides. (Consider the numbers as exact.)

12. Make an attempt to solve the following triangle: $a = 68$, $b = 37$, $c = 29$. Explain.

13. Find the diameter of the largest circular target that can be cut from a triangular piece of cardboard with dimensions 50 cm, 60 cm, 70 cm. (Consider the numbers as exact.)

ANSWERS*

EXERCISE 1, Pages 5–6

1. 5, 9, $\sqrt{10}$, 3
2. 25, 6, $\sqrt{29}$, 4
3. 17, 1, $\sqrt{85}$, 5
5. -7, $\sqrt{17}$, 0
6. -15, $-\sqrt{3}$, -3
7. -3, -4, -2
9. (a) I and II
10. (b) II and III
11. 0, 0
13. $\sqrt{106}$
14. $\sqrt{a^2 + 10a + 41}$

EXERCISE 2, Page 7

1. 490°, $-230°$
2. 680°, $-40°$
3. 400°, $-320°$
5. 90°, $-270°$
6. 20°, $-340°$
7. 100°, $-260°$
9. 281°
10. 111°
11. 196°
13. 30°
14. 240°
15. 299°
17. 7800°

EXERCISE 3, Pages 15–17

1. -0.77, -0.64, 1.2
2. 0.34, 0.94, 0.36
3. 0.17, -0.98, -0.18
5. 0.98, -0.17, -5.7

---

*With a few exceptions, answers are given to all problems except those whose numbers are multiples of four.

6. $-0.98, 0.17, -5.7$

7. $-0.94, -0.34, 2.7$

9. $-0.17, 0.98, -0.18$

10. $0.77, -0.64, -1.2$

11. $0.64, 0.77, 0.84$

13. $0, 0.17, 0.34, 0.50, 0.64, 0.77, 0.87, 0.94, 0.98, 1$

14. $1, 0.98, 0.94, 0.87, 0.77, 0.64, 0.50, 0.34, 0.17, 0$

15. $0, 1, 0$, does not exist, 1, does not exist

17. $-1, 0$, does not exist, 0, does not exist, -1

18. $0, -1, 0$, does not exist, -1, does not exist

19. $-\frac{7}{25}, \frac{24}{25}, -\frac{7}{24}, -\frac{24}{7}, \frac{25}{24}, -\frac{25}{7}$

21. $-\frac{5}{13}, -\frac{12}{13}, \frac{5}{12}, \frac{12}{5}, -\frac{13}{12}, -\frac{13}{5}$

22. $\frac{4}{5}, \frac{3}{5}, \frac{4}{3}, \frac{3}{4}, \frac{5}{3}, \frac{5}{4}$

23. $\dfrac{9\sqrt{85}}{85}, -\dfrac{2\sqrt{85}}{85}, -\dfrac{9}{2}, -\dfrac{2}{9}, -\dfrac{\sqrt{85}}{2}, \dfrac{\sqrt{85}}{9}$

25. $\dfrac{4\sqrt{17}}{17}, \dfrac{\sqrt{17}}{17}, 4, \dfrac{1}{4}, \sqrt{17}, \dfrac{\sqrt{17}}{4}$

26. $-\dfrac{\sqrt{65}}{65}, -\dfrac{8\sqrt{65}}{65}, \dfrac{1}{8}, 8, -\dfrac{\sqrt{65}}{8}, -\sqrt{65}$

27. $-\dfrac{3}{8}, -\dfrac{\sqrt{55}}{8}, \dfrac{3\sqrt{55}}{55}, \dfrac{\sqrt{55}}{3}, -\dfrac{8\sqrt{55}}{55}, -\dfrac{8}{3}$

29. $\dfrac{\sqrt{6}}{3}, -\dfrac{\sqrt{3}}{3}, -\sqrt{2}, -\dfrac{\sqrt{2}}{2}, -\sqrt{3}, \dfrac{\sqrt{6}}{2}$

30. $-\dfrac{4\sqrt{2}}{9}, \dfrac{7}{9}, -\dfrac{4\sqrt{2}}{7}, -\dfrac{7\sqrt{2}}{8}, \dfrac{9}{7}, -\dfrac{9\sqrt{2}}{8}$

31. II

33. IV

34. II

35. III

37. III

38. IV

39. Possible

41. Possible

42. Possible

43. Impossible

45. Impossible

46. Impossible

47. Close to $90°$

49. Close to $0°$

50. Close to $90°$

51. Close to $0°$

53. III

54. II

55. IV

EXERCISE 4, Pages 18–19

1. $\sin\theta = -\dfrac{4\sqrt{3}}{7}$, $\tan\theta = -4\sqrt{3}$, $\cot\theta = -\dfrac{\sqrt{3}}{12}$,

$$\sec\theta = 7,\ \csc\theta = -\frac{7\sqrt{3}}{12}$$

2. $\sin\theta = \dfrac{5\sqrt{26}}{26}$, $\cos\theta = -\dfrac{\sqrt{26}}{26}$, $\cot\theta = -\frac{1}{5}$,

$\sec\theta = -\sqrt{26}$, $\csc\theta = \dfrac{\sqrt{26}}{5}$

3. $\cos\theta = \dfrac{\sqrt{15}}{8}$, $\tan\theta = \dfrac{7\sqrt{15}}{15}$, $\cot\theta = \dfrac{\sqrt{15}}{7}$,

$\sec\theta = \dfrac{8\sqrt{15}}{15}$, $\csc\theta = \frac{8}{7}$

5. $\sin\theta = \frac{15}{17}$, $\cos\theta = \frac{8}{17}$, $\cot\theta = \frac{8}{15}$,
$\sec\theta = \frac{17}{8}$, $\csc\theta = \frac{17}{15}$

6. $\cos\theta = \frac{12}{13}$, $\tan\theta = -\frac{5}{12}$, $\cot\theta = -\frac{12}{5}$,
$\sec\theta = \frac{13}{12}$, $\csc\theta = -\frac{13}{5}$

7. $\sin\theta = \frac{9}{41}$, $\tan\theta = -\frac{9}{40}$, $\cot\theta = -\frac{40}{9}$,
$\sec\theta = -\frac{41}{40}$, $\csc\theta = \frac{41}{9}$

9. $\cos\theta = -\dfrac{3\sqrt{13}}{13}$, $\tan\theta = \frac{2}{3}$, $\cot\theta = \frac{3}{2}$,

$\sec\theta = -\dfrac{\sqrt{13}}{3}$, $\csc\theta = -\dfrac{\sqrt{13}}{2}$

10. $\sin\theta = \frac{2}{9}$, $\tan\theta = \dfrac{2\sqrt{77}}{77}$, $\cot\theta = \dfrac{\sqrt{77}}{2}$

$\sec\theta = \dfrac{9\sqrt{77}}{77}$, $\csc\theta = \frac{9}{2}$

11. $\sin\theta = -\frac{2}{7}$, $\cos\theta = -\dfrac{3\sqrt{5}}{7}$, $\cot\theta = \dfrac{3\sqrt{5}}{2}$,

$\sec\theta = -\dfrac{7\sqrt{5}}{15}$, $\csc\theta = -\frac{7}{2}$

13. $\sin\theta = \frac{1}{9}$, $\cos\theta = -\dfrac{4\sqrt{5}}{9}$, $\tan\theta = -\dfrac{\sqrt{5}}{20}$,

$\cot\theta = -4\sqrt{5}$, $\sec\theta = -\dfrac{9\sqrt{5}}{20}$

14. $\sin\theta = -\dfrac{\sqrt{5}}{5}$, $\cos\theta = -\dfrac{2\sqrt{5}}{5}$, $\tan\theta = \frac{1}{2}$,

$\sec\theta = -\dfrac{\sqrt{5}}{2}$, $\csc\theta = -\sqrt{5}$

15. $\sin \theta = -\dfrac{\sqrt{b^2 - 1}}{b}, \ \cos \theta = \dfrac{1}{b}, \ \tan \theta = -\sqrt{b^2 - 1},$

$\cot \theta = -\dfrac{\sqrt{b^2 - 1}}{b^2 - 1}, \ \csc \theta = -\dfrac{b\sqrt{b^2 - 1}}{b^2 - 1}$

EXERCISE 6, Page 25

1. $\frac{1}{32}$ **2.** $\frac{7}{8}$ **3.** $\frac{9}{16}$ **5.** $\frac{1}{2}$
6. $\frac{3}{4}$ **7.** $\sqrt{6}$ **9.** False   **10.** True
11. False   **13.** True   **14.** False   **15.** True

EXERCISE 7, Page 28

1. (*a*) 6.372, (*b*) 6.37, (*c*) 6.4
2. (*a*) 0.8295, (*b*) 0.829, (*c*) 0.83
3. (*a*) 0.01907, (*b*) 0.0191, (*c*) 0.019
5. (*a*) 21.25°, (*b*) 21.2°, (*c*) 21°
6. (*a*) 38.16°, (*b*) 38.2°, (*c*) 38°
7. (*a*) 44.74°, (*b*) 44.7°, (*c*) 45°
9. 751.5 to 752.5, inclusive
10. 2.485 to 2.495, exclusive
11. 870.5 to 871.5, exclusive
13. 0.91585 to 0.91595, exclusive
14. 4931.5 to 4932.5, inclusive
15. 6.3995 to 6.4005, inclusive

EXERCISE 8, Page 32

1. 0.5704   **2.** 0.9874   **3.** 0.1157   **5.** 0.5990
6. 0.5592   **7.** 2.344   **9.** 3.024   **10.** 0.9861
11. 0.7638   **13.** 45.6°   **14.** 86.4°   **15.** 19.2°
17. 4.0°   **18.** 32.7°   **19.** 77.5°   **21.** 72.1°
22. 50.8°   **23.** 21.9°

EXERCISE 9, Pages 34–35

(The first answer is based on Table 1. If a calculator produces a different result, it is listed in italics in parentheses.)

1. 0.6661	**2.** 0.2161	**3.** 0.6761 (*0.6760*)
5. 2.393	**6.** 4.131 (*4.130*)	**7.** 0.7824
9. 0.4996 (*0.4995*)	**10.** 0.6310	**11.** 0.6778
13. 6.872	**14.** 0.9147	**15.** 0.9530
17. 16.85°	**18.** 88.62°	**19.** 37.29°
21. 42.42° (*42.41°*)	**22.** 79.37° (*79.36°*)	**23.** 8.70°
25. 71.64°	**26.** 41.62°	**27.** 61.28°
29. 64.11° (*64.12°*)	**30.** 27.36° (*27.37°*)	**31.** 73.74°

EXERCISE 10, Pages 41–42

(These answers were obtained by using a calculator. If a table is used, results may vary slightly.)

1. $B = 33.9°$, $b = 112$, $c = 200$
2. $A = 68.9°$, $a = 84.0$, $c = 90.0$
3. $A = 48°$, $b = 3.3$, $c = 5.0$
5. $A = 50.20°$, $a = 25.52$, $b = 21.26$
6. $B = 17.4°$, $a = 582$, $b = 182$
7. $A = 69.4°$, $B = 20.6°$, $c = 574$
9. $A = 49°$, $B = 41°$, $a = 46$
10. $A = 74.29°$, $B = 15.71°$, $b = 536.2$
11. $A = 63.81°$, $B = 26.19°$, $a = 503.4$
13. $A = 23.50°$, $B = 66.50°$, $c = 6019$
14. $A = 67°$, $B = 23°$, $c = 78$
15. $B = 8°$, $a = 98$, $b = 14$
17. $A = 18.4°$, $B = 71.6°$, $b = 7.59$
18. $B = 22.20°$, $b = 0.3681$, $c = 0.9742$
19. $A = 72.58°$, $a = 7474$, $c = 7833$

EXERCISE 11, Pages 44–47

1. 300 meters	**2.** 620 ft
3. 640 ft	**5.** 250 mi, 170 mi

6. N 59° W, S 59° E, 210 mi
9. 233 meters
11. 14 ft
15. 16.4 cm
18. 132.5 meters, 254.8 meters
21. 55°

23. $h \cot \theta \cos \phi$

26. 12 km, 15 km

7. 180 mi, 150 mi
10. 40.15 meters, 114.7 meters
13. 80 ft
17. 6.272 meters, 6.283 meters
19. 2:42 P.M.
22. 29°

25. $\dfrac{a}{\cot \theta + \cot \phi}$

EXERCISE 12, Pages 50–52

1. 1:46 P.M.
3. 21°
6. 289.7°, 57.4 km/h
9. 313.5°, 202 mi/h
11. N 12.7° E, 585 meters per min
13. 9.5°
15. (a) 5415 lb (b) 1121 lb
18. 29 mi/h

2. 10:40 A.M.
5. 198 km/h, 190 km/h
7. 27.5 mi/h, 34.0 mi/h
10. 3.00 min

14. 34.6°, 243 kg
17. (a) 2.2 kg, (b) 0.75 kg
19. (a) 40 lb, (b) 98 lb

EXERCISE 13, Pages 57–58

1. $\cot 33°$
6. $\sin 250°$
11. False
17. False
22. True
27. True

2. $\tan 7A$
7. $\cos 340°$
13. True
18. True
23. False
29. True

3. $\cot (A + B)$
9. $\cot 160°$
14. False
19. False
25. False
30. False

5. $\pm \tan A$
10. $\cos 77°$
15. True
21. False
26. False
31. True

EXERCISE 14, Pages 60–61

1. 1

3. $\dfrac{\cot \theta + 5}{\cot \theta - 7}$

2. $\csc^2 \theta - \csc \theta \cot \theta + \cot^2 \theta$

5. $\pm \cos \theta$

6. 0   **7.** $\sin \theta$

9. $\sec 224° = -\sqrt{1 + \tan^2 224°}$ **10.** $\tan 98° = \dfrac{1}{\cot 98°}$

11. $\cos 112° = -\sqrt{1 - \sin^2 112°}$ **13.** $\cos 5A = \dfrac{1}{\sec 5A}$

14. $\sin 7B = \pm\sqrt{1 - \cos^2 7B}$

15. $\sin \theta = \pm\dfrac{\tan \theta}{\sqrt{1 + \tan^2 \theta}}$, $\cos \theta = \pm\dfrac{1}{\sqrt{1 + \tan^2 \theta}}$, $\cot \theta = \dfrac{1}{\tan \theta}$,

$\sec \theta = \pm\sqrt{1 + \tan^2 \theta}$, $\csc \theta = \pm\dfrac{\sqrt{1 + \tan^2 \theta}}{\tan \theta}$

EXERCISE 15, Pages 67–71

33. Not an identity. Left side $= (4 \cos C)/\sin^2 C \neq 4 \cos C$. True only if $\sin^2 C = 1$, that is, $C = 90°, 270°$, and angles coterminal with them.
34. Not an identity. Left side $= 12 \neq 6$. Not true for any value of θ.
35. An identity.   **37.** An identity.
38. An identity.
39. Not an identity. Left side $= 2 \sec^2 \theta \neq 2$. True only if $\sec^2 \theta = 1$, that is, $\theta = 0°, 180°$, and angles coterminal with them.
45. $0°, 90°, 180°, 270°$ **46.** $0°, 90°, 180°$
47. $0°, 45°, 90°, 180°, 225°, 270°$
65. $a = 12, b = -14, c = 5$ **66.** $C = \frac{3}{40}$

EXERCISE 16, Pages 75–76

1. $\sin 210° = -\frac{1}{2}$, $\cos 210° = -\dfrac{\sqrt{3}}{2}$

2. $\sin 135° = \dfrac{\sqrt{2}}{2}$, $\cos 135° = -\dfrac{\sqrt{2}}{2}$

3. $\sin 315° = -\dfrac{\sqrt{2}}{2}$, $\cos 315° = \dfrac{\sqrt{2}}{2}$

5. $\sin 300° = -\dfrac{\sqrt{3}}{2}$, $\cos 300° = \frac{1}{2}$

6. $\sin 240° = -\dfrac{\sqrt{3}}{2}$, $\cos 240° = -\frac{1}{2}$

7. $\sin 150° = \frac{1}{2}$, $\cos 150° = -\dfrac{\sqrt{3}}{2}$

9. $\sin 855° = \dfrac{\sqrt{2}}{2}$, $\cos 855° = -\dfrac{\sqrt{2}}{2}$

10. $\sin 1050° = -\frac{1}{2}$, $\cos 1050° = \dfrac{\sqrt{3}}{2}$

11. $\sin 600° = -\dfrac{\sqrt{3}}{2}$, $\cos 600° = -\frac{1}{2}$

13. True	**14.** True	**15.** False
17. False	**18.** True	**19.** True
21. 82°, 98°, 278°	**22.** 21°, 159°, 201°	**23.** 132°, 228°, 312°
25. 1.664	**26.** −1.192	**27.** −0.1228
29. 0.9259	**30.** 0.9673	**31.** −0.2385
33. −0.8107	**34.** −0.9968	**35.** −0.3214

EXERCISE 17, Page 78

1. $\sin(-45°) = -\dfrac{\sqrt{2}}{2}$, $\cos(-45°) = \dfrac{\sqrt{2}}{2}$, $\tan(-45°) = -1$

2. $\sin(-60°) = -\dfrac{\sqrt{3}}{2}$, $\cos(-60°) = \frac{1}{2}$, $\tan(-60°) = -\sqrt{3}$

3. $\sin(-90°) = -1$, $\cos(-90°) = 0$, $\tan(-90°)$ does not exist

5. False	**6.** False	**7.** True	**9.** True
10. False	**11.** False	**13.** False	**14.** True
15. False			

18. (*a*), (*e*), (*g*) even; (*b*), (*d*), (*f*) odd; (*c*), (*h*) neither.

EXERCISE 18, Pages 84–85

1. 47.242°	**2.** 83.872°	**3.** 29.833°	**5.** 54°
6. 50°	**7.** −9°	**9.** 170°	**10.** 144°
11. 280°	**13.** 40°	**14.** 1080°	**15.** $22\frac{1}{2}°$

17. $360.96°$ **18.** $-51.57°$ **19.** $57.87°$ **21.** $\dfrac{11\pi}{6}$

22. $\dfrac{5\pi}{4}$ **23.** $\dfrac{2\pi}{3}$ **25.** $\dfrac{\pi}{4}$ **26.** $\dfrac{5\pi}{6}$

27. $\dfrac{7\pi}{6}$ **29.** 10π **30.** $\dfrac{20\pi}{9}$ **31.** $\dfrac{11\pi}{4}$

33. 0.8814 **34.** 0.5817 **35.** 0.1585 **37.** $\tfrac{1}{2}$

38. $\tfrac{1}{2}$ **39.** $\dfrac{\sqrt{2}}{2}$ **41.** $-\tfrac{1}{2}$ **42.** $-\dfrac{\sqrt{2}}{2}$

43. $\tfrac{1}{2}$ **45.** -1 **46.** $\dfrac{\sqrt{3}}{3}$ **47.** $-\dfrac{2\sqrt{3}}{3}$

49. -1 **50.** -1 **51.** 0 **53.** 0.2233

54. 0.5139 **55.** 0.6730 **57.** $\dfrac{5\pi}{6}$, 18π **58.** $\dfrac{7\pi}{30}$

59. $\dfrac{5\pi}{6}$, $\dfrac{\pi}{4}$, $\dfrac{5\pi}{8}$

EXERCISE 19, Pages 90–91

1. $\tfrac{1}{2}$ radian, $28.6°$ **2.** (*a*) 12.0 cm, (*b*) 25.1 cm, (*c*) 96.1 cm
3. (*a*) 40.0 in, (*b*) 8.06 meters
5. (*a*) 14.7 ft, (*b*) 66.0 ft, (*c*) 2110 ft
6. 99.2 cm   **7.** 2700 mi, 3600 mi
9. $38°$ N   **10.** 5200 mi
11. $40°$ N   **13.** 4800 mi
14. $100°$ E   **15.** 520 ft
17. 2.52 km   **18.** moon $0.516°$, sun $0.532°$
21. 10.1 cm   **22.** 15 cm

EXERCISE 20, Pages 93–94

1. $\dfrac{150}{\pi}$ rpm $\rightarrow$ 47.7 rpm

2. 40π radians/sec $\rightarrow$ 126 radians/sec

3. $\dfrac{480}{\pi}$ cm $\rightarrow$ 153 cm   **5.** 806 mi/h

6. $\dfrac{28\pi}{5}$ in/min $\rightarrow$ 17.6 in/min

7. 24π km/h $\rightarrow$ 75.4 km/h

9. $\dfrac{484}{7\pi}$ in $\rightarrow$ 22.0 in

10. 12 cm, 8 cm

11. $\dfrac{24\pi}{25}$ meters/sec $\rightarrow$ 3.02 meters/sec, 96 rpm

EXERCISE 21, Pages 103–105

9. $180°$

11. $60°, 120°$

13. $150°, 210°$

14. $210°, 330°$

15. $135°, 225°$

17. $45°, 135°$

18. $90°, 270°$

19. $30°, 150°$

21. $120°, 300°$

22. $45°, 315°$

23. π

25. $0, \pi$

26. $\dfrac{\pi}{3}, \dfrac{5\pi}{3}$

27. $\dfrac{\pi}{3}, \dfrac{4\pi}{3}$

29. No solution

30. $\dfrac{3\pi}{4}, \dfrac{7\pi}{4}$

31. $132°, 228°$

33. $1°, 181°$

34. $96°, 276°$

35. $39°, 219°$

37. $194°, 346°$

38. $58°, 302°$

39. (a) 21, 19; (b) 11, -3; (c) $-4, -5$

EXERCISE 22, Pages 111–114

1. $\sin 195° = \dfrac{\sqrt{2} - \sqrt{6}}{4} \doteq -0.259$

2. $\cos 345° = \dfrac{\sqrt{6} + \sqrt{2}}{4} \doteq 0.966$

3. $\sin 285° = -\dfrac{\sqrt{6} + \sqrt{2}}{4} \doteq -0.966$

6. 1

7. $-\cos 3\theta$

9. 0

17. $\cos 77° = \cos 11° \cos 66° - \sin 11° \sin 66°$

21. (a) $-\frac{36}{85}$, (b) $-\frac{77}{85}$, (c) Q III

22. (a) $-\frac{4}{5}$, (b) $\frac{3}{5}$, (c) Q IV

23. $\frac{525}{533}$

EXERCISE 23, Pages 116–119

1. $\cos 285° = \dfrac{\sqrt{6} - \sqrt{2}}{4} \doteq 0.259$

2. $\sin 75° = \dfrac{\sqrt{2} + \sqrt{6}}{4} \doteq 0.966$

3. $\tan 105° = -2 - \sqrt{3} \doteq -3.732$

5. $\sqrt{3}$ **6.** $\tan 9\theta$ **7.** $\cos 2\theta$

13. (a) $-\frac{253}{325}$, (b) $-\frac{204}{325}$, (c) $-\frac{323}{36}$, (d) $\frac{253}{204}$

14. (a) $-\frac{672}{697}$, (b) $-\frac{185}{697}$, (c) $\frac{528}{455}$, (d) $\frac{672}{185}$

17. True   **18.** False   **19.** True

21. $\cos 23° = \cos 79° \cos 56° + \sin 79° \sin 56°$

22. $\sin 61° = \sin 88° \cos 27° - \cos 88° \sin 27°$

23. $\tan 12° = \dfrac{\tan 36° - \tan 24°}{1 + \tan 36° \tan 24°}$

41. $\sqrt{2} \sin (\theta + 315°) = \sqrt{2} \sin (\theta + 7\pi/4)$

42. $13 \sin (\theta + 157.4°) = 13 \sin (\theta + 2.747)$

43. $25 \sin (\theta + 16.3°) = 25 \sin (\theta + 0.284)$

45. $-50\sqrt{3} \sin \theta - 50 \cos \theta$

EXERCISE 24, Pages 123–126

3. $\sin 67\frac{1}{2}° = \frac{1}{2}\sqrt{2 + \sqrt{2}}$, $\cos 67\frac{1}{2}° = \frac{1}{2}\sqrt{2 - \sqrt{2}}$

5. $\pm \sin 9\theta$ **6.** $\cos 160°$ **7.** $8 \cos 2B$

9. $\frac{1}{2} \sin \dfrac{\theta}{3}$ **10.** $-\cos 2A$ **11.** $\cos^2 5C$

13. $\cos \dfrac{A}{2} = -\dfrac{\sqrt{35}}{7}$, $\cos 2A = -\frac{31}{49}$

14. $\sin \dfrac{B}{2} = \dfrac{3\sqrt{13}}{13}$, $\sin 2B = \frac{120}{169}$

15. $\sin 5A = \dfrac{\sqrt{10}}{10}$, $\sin 20A = \frac{24}{25}$

17. $\sin 128° = 2 \sin 64° \cos 64°$

18. $\cos 246° = 1 - 2 \sin^2 123°$

19. $\cos 130° = -\sqrt{\dfrac{1 + \cos 260°}{2}}$

21. $\sin 3A = \pm\sqrt{\dfrac{1 - \cos 6A}{2}}$

22. $\cos 7\theta = \pm\sqrt{\dfrac{1 + \cos 14\theta}{2}}$

23. $\cos 6C = 2\cos^2 3C - 1$

25. True   **26.** True   **27.** True
29. True   **30.** True   **31.** False
33. False   **34.** False   **35.** True

EXERCISE 25, Pages 128–129

1. $\frac{1}{2}\cos 10\theta + \frac{1}{2}\cos 4\theta$

2. $6\sin 17A + 6\sin A$

3. $\cos \theta - \cos 2\theta$

5. $4\sin 149° - 4\sin 65°$

6. $\frac{1}{2}\cos 175° - \frac{1}{2}\cos 237°$

7. $-\sqrt{3}\cos 50°$

9. $-2\sin 12A \sin 3A$

10. $\sqrt{3}\sin (B + 30°)$

EXERCISE 26, Pages 134–137

1. $45°, 135°, 225°, 315°$

2. $30°, 150°, 210°, 330°$

3. $60°, 120°, 240°, 300°$

5. $247°, 307°$

6. $35°, 305°$

7. $45°, 135°, 150°, 210°$

9. $30°, 150°, 330°$

10. $60°, 240°, 300°$

11. $0°, 45°, 180°, 225°$

13. $0, \dfrac{2\pi}{3}, \dfrac{4\pi}{3}$

14. $\dfrac{2\pi}{3}, \dfrac{4\pi}{3}, \dfrac{3\pi}{2}$

15. $\dfrac{7\pi}{6}, \dfrac{3\pi}{2}, \dfrac{11\pi}{6}$

17. $\dfrac{\pi}{6}, \dfrac{7\pi}{6}$

18. $0, \dfrac{\pi}{3}, \dfrac{2\pi}{3}, \pi$

19. No solution

21. No solution

22. No solution

23. $0, \dfrac{\pi}{6}, \dfrac{5\pi}{6}, \pi$

25. $0, \dfrac{3\pi}{4}, \pi, \dfrac{5\pi}{4}$ **26.** $\dfrac{\pi}{2}, \dfrac{3\pi}{2}$

27. $0, \dfrac{2\pi}{3}$

29. $30°, 90°, 150°, 210°, 270°, 330°$

30. $90°, 210°, 330°$ **31.** $270°$

33. $60°, 300°$ **34.** $60°, 120°, 240°, 300°$

35. The given equation is an identity. It holds true for all permissible values of θ, that is, all θ except $0°$, $90°$, $180°$, $270°$, and angles coterminal with them.

37. $225°$ **38.** $150°$

39. $270°$ **41.** $78.5°, 80.4°, 279.6°, 281.5°$

42. $63.4°, 108.4°, 243.4°, 288.4°$

43. $19.5°, 160.5°, 199.5°, 340.5°$

45. $33.7°, 213.7°$ **46.** $229.8°, 310.2°$

47. $45°, 76.0°, 225°, 256.0°$ **49.** $113.1°, 353.1°$

50. $232.6°, 352.6°$ **51.** $36.8°, 110.6°$

53. $20°, 110°, 140°, 200°, 290°, 320°$

54. $35°, 95°, 155°, 165°, 215°, 275°, 335°, 345°$

55. $30°, 90°, 150°, 210°, 270°, 330°$

57. $30°, 60°, 210°, 240°$

58. $90°, 270°$ **59.** $60°, 300°$

61. $135°, 315°$ **62.** $300°$

EXERCISE 27, Pages 145–147

1. $\dfrac{2\pi}{5}, 1$ **2.** $\dfrac{\pi}{2}, 3$ **3.** $2, \frac{1}{2}$

5. $8\pi, 7$ **6.** $\frac{6}{5}, \infty$ **7.** $\dfrac{5\pi}{4}, \infty$

9. $\dfrac{\pi}{2}, 3.5, 1$ unit left   **10.** $\dfrac{4\pi}{3}, 8, \dfrac{5\pi}{2}$ units right

11. $6, 1, 6$ units right

EXERCISE 28, Pages 150–151

1. 0 **2.** $\frac{1}{4}$ **3.** 3

5. 6 **6.** 1 **7.** $-\frac{3}{5}$

9. $\frac{2}{3}$ **10.** $-\frac{4}{3}$ **11.** 0

13. -2 **14.** 4   **15.** $\frac{3}{4}$

17. $\frac{1}{6}$ **18.** 1   **19.** 15

21. 9   **22.** 3   **23.** -4

25. $\frac{1}{64}$ **26.** 31   **27.** 4

29. $\log_9 243 = \frac{5}{2}$ **30.** $\log_{81} \frac{1}{9} = -\frac{1}{2}$ **31.** $\log_{64} \frac{1}{2} = -\frac{1}{6}$

33. $b^c = a$ **34.** $1024^{2/5} = 16$ **35.** $121^{1/2} = 11$

37. True   **38.** True   **39.** True

EXERCISE 29, Pages 153–154

1. 1.15   **2.** -0.55 **3.** 0.18

5. 0.24   **6.** 1.44   **7.** 1.50

9. 2.15   **10.** 0.44   **11.** 1.09

13. $\log \dfrac{a + b}{b}$ **14.** $\log P(1 + i)^n$ **15.** $\log a^6 \sqrt{b^2 + c^2}$

17. $\log x^4 \sqrt[6]{y^2 z}$ **18.** $\log r \sqrt{s/t}$ **19.** $\log (x + y)z$

21. False   **22.** True   **23.** False

25. True   **26.** False   **27.** True

29. True   **30.** True   **31.** True

33. False   **34.** True

EXERCISE 30, Page 157

1. -1 **2.** No solution   **3.** $3 \pm \sqrt{2}$

5. 4   **6.** 11   **7.** 20

9. $x = 4 + \sqrt{16 + b^y}$

10. $x = \dfrac{fb^{2y} - a}{c - gb^{2y}}$

EXERCISE 31, Pages 163–164

1. $B = 68.8°$, $a = 437$, $b = 601$

2. $A = 33°$, $b = 3.5$, $c = 2.3$

3. $B = 79.6°$, $a = 52.5$, $c = 24.2$

5. $A = 42.4°$, $b = 4.57$, $c = 10.6$

6. $C = 43.1°$, $a = 0.210$, $c = 0.143$

The answers listed for Exercises 31 to 35 were obtained by using a calculator. If tables are used, results may vary slightly.

7. $A = 47.9°$, $a = 1130$, $b = 1520$
9. $A = 30.20°$, $a = 0.001884$, $c = 0.003725$
10. $C = 100.00°$, $a = 1564$, $b = 6487$
11. $C = 77.83°$, $b = 15.44$, $c = 15.90$
13. 1.6 mi, 1.0 mi
14. 350 km, 620 km
15. 105 meters/min, 406 meters/min

EXERCISE 32, Page 169

1. $B = 74.6°$, $C = 65.4°$, $c = 28.3$
$B' = 105.4°$, $C' = 34.6°$, $c' = 17.7$
2. No triangle
3. $B = 59.2°$, $C = 41.7°$, $c = 0.347$
5. No triangle
6. $B = 90.0°$, $C = 60.0°$, $c = 68.8$
7. $B = 43.3°$, $C = 109.3°$, $c = 3.55$
$B' = 136.7°$, $C' = 15.9°$, $c' = 1.03$
9. $B = 18.89°$, $C = 48.76°$, $b = 3211$
10. $B = 129.18°$, $C = 32.24°$, $b = 9.463$
$B' = 13.66°$, $C' = 147.76°$, $b' = 2.882$
11. No triangle
13. There are two Charlestons that satisfy the conditions of the problem. Charleston, W.Va., is 120 mi from Zanesville; Charleston, S.C., is 510 mi from Zanesville.
14. 13°
15. 139,000,000 mi or 30,000,000 mi

EXERCISE 33, Pages 175–176

1. $A = 27.7°$, $B = 40.5°$, $C = 111.8°$
2. $A = 28°$, $B = 31°$, $C = 121°$

3. $A = 120.0°$, $B = 43.6°$, $C = 16.4°$
5. $a = 93$, $B = 56°$, $C = 18°$
6. $b = 7.4$, $A = 10°$, $C = 108°$
7. $c = 849$, $A = 36.6°$, $B = 46.4°$
9. $A = 38.20°$, $B = 59.99°$, $C = 81.80°$
10. $c = 6765$, $A = 71.37°$, $B = 48.97°$
11. $b = 538$, $A = 22.5°$, $C = 132.4°$
13. 2.8 meters (exactly)
14. N 63° E or S 37° W
15. S 73° E, N 43° E
17. 20°, 80° or 160°, 100°
18. N 76° E, 16 km/h

EXERCISE 34, Pages 178–179

1. 110   **2.** 12.1   **3.** 4.3
5. 8400 (exactly)   **6.** $24\sqrt{29} \doteq 129.2$
7. 23.4   **9.** 787,000   **10.** 12.18   **11.** 1822
14. 13.4 meters

EXERCISE 35, Pages 179–182

1. 11
3. 72° or 108°
6. 47.2° or 132.8°
9. 337 meters or 383 meters
11. 1010 meters
14. 574.2 meters
17. 380 mi, 260 mi
18. Two cities named Columbus satisfy the conditions of the problem. Columbus, Ohio, is 150 mi from Aliquippa; Columbus, Ind., is 320 mi from Aliquippa.
19. N 49° E, 150 mi
21. 68°
23. (a) 8°, (b) 172°, (c) 1 h 2 min
26. 14°, 64°
29. 40 ft (exactly)
30. $x = \dfrac{m \sin(\alpha + \gamma) \sin(\alpha + \beta)}{\sin(\beta - \gamma) \cos \alpha}$ (meters)
33. $\frac{1}{2}\sqrt{70}$

2. 1040
5. 71.6°
7. 593
10. N 27° W
13. N 74° W
15. $\sqrt{650} \doteq 25.5$

22. 252°
25. 279 km/h, 279.0°
27. 51° or 29°

EXERCISE 36, Page 188

1. 0

2. $\dfrac{5\pi}{6}$

3. $-\dfrac{\pi}{2}$

5. $-\dfrac{\pi}{4}$

6. $\dfrac{\pi}{2}$

7. $\dfrac{\pi}{4}$

9. $\dfrac{\pi}{6}$

10. $\dfrac{\pi}{3}$

11. $\dfrac{\pi}{3}$

13. 0

14. $\dfrac{\pi}{6}$

15. $\dfrac{2\pi}{3}$

17. $-\dfrac{\pi}{3}$

18. $\dfrac{\pi}{4}$

19. $-\dfrac{\pi}{6}$

21. $\dfrac{\pi}{4}$

22. -0.3910

23. 1.3788

25. 2.4836

26. 1.4556

27. 1.1816

EXERCISE 37, Pages 192–194

1. $\dfrac{1}{u}$

2. $\sqrt{u^2 + 1}$

3. $\dfrac{u}{\sqrt{1 - u^2}}$

5. $\frac{12}{13}$

6. $\dfrac{\sqrt{35}}{6}$

7. $-\frac{4}{3}$

9. $v\sqrt{1 - u^2} - u\sqrt{1 - v^2}$

10. $u\sqrt{1 - v^2} + v\sqrt{1 - u^2}$

11. $v\sqrt{1 - u^2} - u\sqrt{1 - v^2}$

13. $\dfrac{3\sqrt{13} - 5}{16}$

14. $-\frac{1}{3}$

15. $\dfrac{v - \sqrt{1 - v^2}}{\sqrt{2}}$

17. $2u\sqrt{1 - u^2}$

18. $1 - 2u^2$

19. $2v^2 - 1$

21. $2v$

22. $4u$

23. $-\dfrac{\pi}{3}$

25. $\dfrac{2\pi}{5}$

26. $-\dfrac{\pi}{7}$

27. $\dfrac{3\pi}{10}$

29. $\dfrac{\pi}{6}$

30. $-\dfrac{\pi}{2}$

31. $u,\ \sqrt{1 - u^2}$

33. $\dfrac{u + 1}{\sqrt{u^2 + 2u + 5}},\ \dfrac{u + 1}{2}$

34. $\dfrac{6}{\sqrt{u^2 - 36}},\ \dfrac{u}{6}$

45. True   **46.** True   **47.** True   **49.** True
50. True   **51.** True   **53.** False if $u < 0$
54. False if $u < 0$
55. False; right side should be $1/\sqrt{1 - u^2}$
57. $\frac{1}{3}$ **58.** $-\frac{1}{2}$ **59.** 0   **61.** $\frac{1}{2}$, 1

62. 1   **63.** $-\dfrac{\sqrt{3}}{2}$

EXERCISE 38, Pages 195–196

1. $g(u) = \dfrac{4u - 3}{2}$

2. $g(u) = \dfrac{6u - 5}{u - 4}$

3. $g(u) = u^2 - 6u + 11$

EXERCISE 39, Page 200

1. $4 + 7i$ **2.** $-9 - 3i$ **3.** $-6 - 3i$
5. $-59 - 13i$ **6.** $11 + 2i\sqrt{3}$ **7.** $23 - 11i\sqrt{7}$
9. $45 - 108i$ **10.** $2i$ **11.** $-i$
13. -1 **14.** 1   **15.** $\frac{30}{17} - \frac{1}{17}i$

17. $\dfrac{13}{11} - \dfrac{8\sqrt{2}}{11}i$ **18.** $6 + 5i$ **19.** $2 + 2i$

21. $x = 3, y = -4$ **22.** $x = 5, y = -19$
23. $x = 6, y = 8;\ x = -8, y = -6$ **25.** 0

EXERCISE 40, Pages 204–205

1. $-3 + 4i$ **2.** $4 - 4i$ **3.** $-5 - 7i$
5. $2 + 4i$ **6.** $-5i$ **7.** $1 + i$

9. $2\sqrt{2}(\cos 315° + i \sin 315°)$ **10.** $8\sqrt{2}(\cos 45° + i \sin 45°)$

11. $2(\cos 225° + i \sin 225°)$ **13.** $9(\cos 180° + i \sin 180°)$

14. $6(\cos 270° + i \sin 270°)$ **15.** $3(\cos 0° + i \sin 0°)$

17. $4(\cos 210° + i \sin 210°)$

18. $5(\cos 323.13° + i \sin 323.13°)$

19. $13(\cos 247.38° + i \sin 247.38°)$

21. $3.759 + 1.368i$ **22.** $-3.420 + 9.397i$

23. $6.928 - 4.000i$

EXERCISE 41, Pages 210–211

1. $6\sqrt{3} - 6i$ **2.** -35 **3.** $-3 + 3i\sqrt{3}$

5. $-28 + 28i$ **6.** $2\sqrt{3} + 2i$ **7.** $2i\sqrt{2}$

9. $12\sqrt{3} + 12i$ **10.** $1 + i\sqrt{3}$ **11.** $-24i$

13. $500 + 500i\sqrt{3}$ **14.** $-16\sqrt{3} - 16i$ **15.** $32i$

17. $-16 - 16i$ **18.** 256 **19.** -1

21. $64 - 64i\sqrt{3}$

22. $28 + 96i$, using a calculator; $28.04 + 95.99i$, using Table 1

23. $-1.792 + 4.668i,\ -3.147 - 3.886i,\ 4.938 - 0.782i$

25. $-3 + 3i,\ 3 - 3i$ **26.** $4 + 4i,\ -4 - 4i$

27. $-2 + 2i\sqrt{3},\ 2 - 2i\sqrt{3}$ **29.** $5 + 5i\sqrt{3},\ -10,\ 5 - 5i\sqrt{3}$

30. $-0.518 + 1.932i,\ -\sqrt{2} - i\sqrt{2},\ 1.932 - 0.518i$

31. $2,\ 2i,\ -2,\ -2i$

33. $1 + i\sqrt{3},\ -\sqrt{3} + i,\ -1 - i\sqrt{3},\ \sqrt{3} - i$

34. $3.827 + 9.239i,\ -9.239 + 3.827i,\ -3.827 - 9.239i,\ 9.239 - 3.827i$

35. $\sqrt{3} + i,\ 2i,\ -\sqrt{3} + i,\ -\sqrt{3} - i,\ -2i,\ \sqrt{3} - i$

37. $3,\ 0.927 + 2.853i,\ -2.427 + 1.763i,\ -2.427 - 1.763i,$
$0.927 - 2.853i$

38. $1,\ 0.809 + 0.588i,\ 0.309 + 0.951i,\ -0.309 + 0.951i,$
$-0.809 + 0.588i,\ -1,\ -0.809 - 0.588i,\ -0.309 - 0.951i,$
$0.309 - 0.951i,\ 0.809 - 0.588i$

EXERCISE 42, Page 214

1. 0.09983342

EXERCISE 43, Pages 221–222

1. 1.9143
2. 5.5647 − 10
3. 5.8681
5. 8.2765 − 10
6. 0.9614
7. 6.4330 − 10
9. 3.8102
10. 4.7716
11. 0.6561
13. 0.000 0712
14. 2900
15. 336
17. 4,700,000
18. 0.829
19. 0.00701
21. 9.55
22. 0.000 00567
23. 18,000

EXERCISE 44, Pages 223–224

1. 7.4817 − 10
2. 1.6734
3. 5.7402 − 10
5. 2.9855
6. 8.3967 − 10
7. 1.9168
9. 9.4658 − 10
10. 6.8052
11. 8.9943 − 10
13. 4.7571
14. 6.8753 − 10
15. 0.4936
17. 0.000 008604
18. 403.7
19. 9245
21. 299,100
22. 0.006409
23. 0.5508
25. 0.000 1878
26. 1.352
27. 1,020,000
29. 34.67
30. 5125
31. 0.000 06781

EXERCISE 45, Pages 226–228

(The first answer is based on Table 2. If a calculator produces a different result, it is listed in italics in parentheses.)

1. 130
2. 0.479
3. 9530 (*9540*)
5. 0.000 489 (*0.000 488*)
6. 50.7
7. 0.0449
9. 23.5
10. 2.09
11. 227
13. 0.736 (*0.737*)
14. 0.0939
15. 0.841
17. −0.00596
18. −0.000 244 (*−0.000 243*)
19. −1.08
21. 1.98
22. 2.84
23. −0.00278
25. 0.0235
26. 0.00841 (*0.00840*)
27. 0.0399
29. 1.107
30. 1667 (*1666*)
31. 3705
33. 2135 (*2137*)
34. 297.1
35. 45,230
37. 0.868
38. 21.30
39. 10.0 (*10.1*)
41. 7567
42. 0.0354

43. 0.032
46. 2.45 lb/in^2

45. 2.031 sec
47. $193,800

EXERCISE 46, Pages 230–231

(The first answer is based on Table 2. If a calculator produces a different result, it is listed in italics in parentheses.)

1. -0.436 (-0.434) **2.** 2.70   **3.** 20.3
5. -0.723 **6.** 14.8

7. $\dfrac{\log c}{\log a + \log b - \log c}$

9. $x = \log_{10}(y \pm \sqrt{y^2 - 1}), y \geq 1$ **10.** 0.833
11. 2.41   **13.** 4.61   **14.** -0.298
15. -2.84 **17.** -1.98 **18.** 2.10
19. 9.02   **21.** (a) 92.2 mg, (b) 259 (258)
22. (a) 3030, (b) 3.169 (3.170)

EXERCISE 47, Page 233

1. $9.9935 - 10$ **2.** $9.9996 - 10$ **3.** 0.0334
5. 0.1901   **6.** 0.2945   **7.** $9.6073 - 10$
9. $9.1516 - 10$ **10.** $9.8640 - 10$ **11.** $9.9790 - 10$
13. $9.7663 - 10$ **14.** $9.3048 - 10$ **15.** 0.5155
17. 29.10°   **18.** 48.60°   **19.** 53.40°
21. 46.88°   **22.** 31.91°   **23.** 17.42°
25. 81.82°   **26.** 5.43°   **27.** 35.07°
29. 10.57°   **30.** 63.18°   **31.** 75.26°

EXERCISE 48, Page 237

(These results are based on Tables 2 and 3 and the law of tangents.)

5. $A = 83.0°, B = 31.0°, c = 448$ or 449, depending upon the method used
6. $A = 83.3°, B = 44.9°, c = 66.1$

7. $A = 8.2°$, $B = 69.2°$, $c = 7.73$ or 7.74
9. $A = 98.2°$, $C = 62.4°$, $b = 24.4$
10. $B = 20.4°$, $C = 33.6°$, $a = 638$ or 639
11. $A = 42.00°$, $C = 62.00°$, $b = 5679$ or 5677
13. $44°$, $112°$
14. 30.9 cm, 69.1 cm

EXERCISE 49, Pages 240–241

(These results are based on Tables 2 and 3 and the half-angle formulas.)

2. $A = 44.4°$, $B = 57.1°$, $C = 78.5°$
3. $A = 43.4°$, $B = 47.9°$, $C = 88.6°$
5. $A = 48.2°$, $B = 16.6°$, $C = 115.1°$
6. $A = 34.90°$, $B = 89.06°$, $C = 56.04°$
7. $A = 51.04°$, $B = 103.26°$, $C = 25.68°$
9. $A = 64.58°$, $B = 104.26°$, $C = 11.14°$
10. 16
11. 4 and 5, exactly
13. $\dfrac{40\sqrt{6}}{3}$ cm $\doteq 32.66$ cm

LOGARITHMIC
AND
TRIGONOMETRIC
TABLES

TABLE 1

Radians	Degrees	sin	tan	cot	cos		
.0000	0.0°	.0000	.0000	—	1.0000	90.0°	1.5708
.0017	0.1°	.0017	.0017	573.0	1.0000	89.9°	1.5691
.0035	0.2°	.0035	.0035	286.5	1.0000	89.8°	1.5673
.0052	0.3°	.0052	.0052	191.0	1.0000	89.7°	1.5656
.0070	0.4°	.0070	.0070	143.2	1.0000	89.6°	1.5638
.0087	0.5°	.0087	.0087	114.6	1.0000	89.5°	1.5621
.0105	0.6°	.0105	.0105	95.49	.9999	89.4°	1.5603
.0122	0.7°	.0122	.0122	81.85	.9999	89.3°	1.5586
.0140	0.8°	.0140	.0140	71.62	.9999	89.2°	1.5568
.0157	0.9°	.0157	.0157	63.66	.9999	89.1°	1.5551
.0175	1.0°	.0175	.0175	57.29	.9998	89.0°	1.5533
.0192	1.1°	.0192	.0192	52.08	.9998	88.9°	1.5516
.0209	1.2°	.0209	.0209	47.74	.9998	88.8°	1.5499
.0227	1.3°	.0227	.0227	44.07	.9997	88.7°	1.5481
.0244	1.4°	.0244	.0244	40.92	.9997	88.6°	1.5464
.0262	1.5°	.0262	.0262	38.19	.9997	88.5°	1.5446
.0279	1.6°	.0279	.0279	35.80	.9996	88.4°	1.5429
.0297	1.7°	.0297	.0297	33.69	.9996	88.3°	1.5411
.0314	1.8°	.0314	.0314	31.82	.9995	88.2°	1.5394
.0332	1.9°	.0332	.0332	30.14	.9995	88.1°	1.5376
.0349	2.0°	.0349	.0349	28.64	.9994	88.0°	1.5359
.0367	2.1°	.0366	.0367	27.27	.9993	87.9°	1.5341
.0384	2.2°	.0384	.0384	26.03	.9993	87.8°	1.5324
.0401	2.3°	.0401	.0402	24.90	.9992	87.7°	1.5307
.0419	2.4°	.0419	.0419	23.86	.9991	87.6°	1.5289
.0436	2.5°	.0436	.0437	22.90	.9990	87.5°	1.5272
.0454	2.6°	.0454	.0454	22.02	.9990	87.4°	1.5254
.0471	2.7°	.0471	.0472	21.20	.9989	87.3°	1.5237
.0489	2.8°	.0488	.0489	20.45	.9988	87.2°	1.5219
.0506	2.9°	.0506	.0507	19.74	.9987	87.1°	1.5202
.0524	3.0°	.0523	.0524	19.08	.9986	87.0°	1.5184
.0541	3.1°	.0541	.0542	18.46	.9985	86.9°	1.5167
.0559	3.2°	.0558	.0559	17.89	.9984	86.8°	1.5149
.0576	3.3°	.0576	.0577	17.34	.9983	86.7°	1.5132
.0593	3.4°	.0593	.0594	16.83	.9982	86.6°	1.5115
.0611	3.5°	.0610	.0612	16.35	.9981	86.5°	1.5097
.0628	3.6°	.0628	.0629	15.89	.9980	86.4°	1.5080
.0646	3.7°	.0645	.0647	15.46	.9979	86.3°	1.5062
.0663	3.8°	.0663	.0664	15.06	.9978	86.2°	1.5045
.0681	3.9°	.0680	.0682	14.67	.9977	86.1°	1.5027
.0698	4.0°	.0698	.0699	14.30	.9976	86.0°	1.5010
.0716	4.1°	.0715	.0717	13.95	.9974	85.9°	1.4992
.0733	4.2°	.0732	.0734	13.62	.9973	85.8°	1.4975
.0750	4.3°	.0750	.0752	13.30	.9972	85.7°	1.4957
.0768	4.4°	.0767	.0769	13.00	.9971	85.6°	1.4940
.0785	4.5°	.0785	.0787	12.71	.9969	85.5°	1.4923
.0803	4.6°	.0802	.0805	12.43	.9968	85.4°	1.4905
.0820	4.7°	.0819	.0822	12.16	.9966	85.3°	1.4888
.0838	4.8°	.0837	.0840	11.91	.9965	85.2°	1.4870
.0855	4.9°	.0854	.0857	11.66	.9963	85.1°	1.4853
.0873	5.0°	.0872	.0875	11.43	.9962	85.0°	1.4835
		cos	cot	tan	sin	Degrees	Radians

TABLE 1

Radians	Degrees	sin	tan	cot	cos		
.0873	5.0°	.0872	.0875	11.43	.9962	85.0°	1.4835
.0890	5.1°	.0889	.0892	11.20	.9960	84.9°	1.4818
.0908	5.2°	.0906	.0910	10.99	.9959	84.8°	1.4800
.0925	5.3°	.0924	.0928	10.78	.9957	84.7°	1.4783
.0942	5.4°	.0941	.0945	10.58	.9956	84.6°	1.4765
.0960	5.5°	.0958	.0963	10.39	.9954	84.5°	1.4748
.0977	5.6°	.0976	.0981	10.20	.9952	84.4°	1.4731
.0995	5.7°	.0993	.0998	10.02	.9951	84.3°	1.4713
.1012	5.8°	.1011	.1016	9.845	.9949	84.2°	1.4696
.1030	5.9°	.1028	.1033	9.677	.9947	84.1°	1.4678
.1047	6.0°	.1045	.1051	9.514	.9945	84.0°	1.4661
.1065	6.1°	.1063	.1069	9.357	.9943	83.9°	1.4643
.1082	6.2°	.1080	.1086	9.205	.9942	83.8°	1.4626
.1100	6.3°	.1097	.1104	9.058	.9940	83.7°	1.4608
.1117	6.4°	.1115	.1122	8.915	.9938	83.6°	1.4591
.1134	6.5°	.1132	.1139	8.777	.9936	83.5°	1.4573
.1152	6.6°	.1149	.1157	8.643	.9934	83.4°	1.4556
.1169	6.7°	.1167	.1175	8.513	.9932	83.3°	1.4539
.1187	6.8°	.1184	.1192	8.386	.9930	83.2°	1.4521
.1204	6.9°	.1201	.1210	8.264	.9928	83.1°	1.4504
.1222	7.0°	.1219	.1228	8.144	.9925	83.0°	1.4486
.1239	7.1°	.1236	.1246	8.028	.9923	82.9°	1.4469
.1257	7.2°	.1253	.1263	7.916	.9921	82.8°	1.4451
.1274	7.3°	.1271	.1281	7.806	.9919	82.7°	1.4434
.1292	7.4°	.1288	.1299	7.700	.9917	82.6°	1.4416
.1309	7.5°	.1305	.1317	7.596	.9914	82.5°	1.4399
.1326	7.6°	.1323	.1334	7.495	.9912	82.4°	1.4382
.1344	7.7°	.1340	.1352	7.396	.9910	82.3°	1.4364
.1361	7.8°	.1357	.1370	7.300	.9907	82.2°	1.4347
.1379	7.9°	.1374	.1388	7.207	.9905	82.1°	1.4329
.1396	8.0°	.1392	.1405	7.115	.9903	82.0°	1.4312
.1414	8.1°	.1409	.1423	7.026	.9900	81.9°	1.4294
.1431	8.2°	.1426	.1441	6.940	.9898	81.8°	1.4277
.1449	8.3°	.1444	.1459	6.855	.9895	81.7°	1.4259
.1466	8.4°	.1461	.1477	6.772	.9893	81.6°	1.4242
.1484	8.5°	.1478	.1495	6.691	.9890	81.5°	1.4224
.1501	8.6°	.1495	.1512	6.612	.9888	81.4°	1.4207
.1518	8.7°	.1513	.1530	6.535	.9885	81.3°	1.4190
.1536	8.8°	.1530	.1548	6.460	.9882	81.2°	1.4172
.1553	8.9°	.1547	.1566	6.386	.9880	81.1°	1.4155
.1571	9.0°	.1564	.1584	6.314	.9877	81.0°	1.4137
.1588	9.1°	.1582	.1602	6.243	.9874	80.9°	1.4120
.1606	9.2°	.1599	.1620	6.174	.9871	80.8°	1.4102
.1623	9.3°	.1616	.1638	6.107	.9869	80.7°	1.4085
.1641	9.4°	.1633	.1655	6.041	.9866	80.6°	1.4067
.1658	9.5°	.1650	.1673	5.976	.9863	80.5°	1.4050
.1676	9.6°	.1668	.1691	5.912	.9860	80.4°	1.4032
.1693	9.7°	.1685	.1709	5.850	.9857	80.3°	1.4015
.1710	9.8°	.1702	.1727	5.789	.9854	80.2°	1.3998
.1728	9.9°	.1719	.1745	5.730	.9851	80.1°	1.3980
.1745	10.0°	.1736	.1763	5.671	.9848	80.0°	1.3963
		cos	cot	tan	sin	Degrees	Radians

TABLE 1

Radians	Degrees	sin	tan	cot	cos		
.1745	10.0°	.1736	.1763	5.671	.9848	80.0°	1.3963
.1763	10.1°	.1754	.1781	5.614	.9845	79.9°	1.3945
.1780	10.2°	.1771	.1799	5.558	.9842	79.8°	1.3928
.1798	10.3°	.1788	.1817	5.503	.9839	79.7°	1.3910
.1815	10.4°	.1805	.1835	5.449	.9836	79.6°	1.3893
.1833	10.5°	.1822	.1853	5.396	.9833	79.5°	1.3875
.1850	10.6°	.1840	.1871	5.343	.9829	79.4°	1.3858
.1868	10.7°	.1857	.1890	5.292	.9826	79.3°	1.3840
.1885	10.8°	.1874	.1908	5.242	.9823	79.2°	1.3823
.1902	10.9°	.1891	.1926	5.193	.9820	79.1°	1.3806
.1920	11.0°	.1908	.1944	5.145	.9816	79.0°	1.3788
.1937	11.1°	.1925	.1962	5.097	.9813	78.9°	1.3771
.1955	11.2°	.1942	.1980	5.050	.9810	78.8°	1.3753
.1972	11.3°	.1959	.1998	5.005	.9806	78.7°	1.3736
.1990	11.4°	.1977	.2016	4.959	.9803	78.6°	1.3718
.2007	11.5°	.1994	.2035	4.915	.9799	78.5°	1.3701
.2025	11.6°	.2011	.2053	4.872	.9796	78.4°	1.3683
.2042	11.7°	.2028	.2071	4.829	.9792	78.3°	1.3666
.2059	11.8°	.2045	.2089	4.787	.9789	78.2°	1.3648
.2077	11.9°	.2062	.2107	4.745	.9785	78.1°	1.3631
.2094	12.0°	.2079	.2126	4.705	.9781	78.0°	1.3614
.2112	12.1°	.2096	.2144	4.665	.9778	77.9°	1.3596
.2129	12.2°	.2113	.2162	4.625	.9774	77.8°	1.3579
.2147	12.3°	.2130	.2180	4.586	.9770	77.7°	1.3561
.2164	12.4°	.2147	.2199	4.548	.9767	77.6°	1.3544
.2182	12.5°	.2164	.2217	4.511	.9763	77.5°	1.3526
.2199	12.6°	.2181	.2235	4.474	.9759	77.4°	1.3509
.2217	12.7°	.2198	.2254	4.437	.9755	77.3°	1.3491
.2234	12.8°	.2215	.2272	4.402	.9751	77.2°	1.3474
.2251	12.9°	.2233	.2290	4.366	.9748	77.1°	1.3456
.2269	13.0°	.2250	.2309	4.331	.9744	77.0°	1.3439
.2286	13.1°	.2267	.2327	4.297	.9740	76.9°	1.3422
.2304	13.2°	.2284	.2345	4.264	.9736	76.8°	1.3404
.2321	13.3°	.2300	.2364	4.230	.9732	76.7°	1.3387
.2339	13.4°	.2317	.2382	4.198	.9728	76.6°	1.3369
.2356	13.5°	.2334	.2401	4.165	.9724	76.5°	1.3352
.2374	13.6°	.2351	.2419	4.134	.9720	76.4°	1.3334
.2391	13.7°	.2368	.2438	4.102	.9715	76.3°	1.3317
.2409	13.8°	.2385	.2456	4.071	.9711	76.2°	1.3299
.2426	13.9°	.2402	.2475	4.041	.9707	76.1°	1.3282
.2443	14.0°	.2419	.2493	4.011	.9703	76.0°	1.3265
.2461	14.1°	.2436	.2512	3.981	.9699	75.9°	1.3247
.2478	14.2°	.2453	.2530	3.952	.9694	75.8°	1.3230
.2496	14.3°	.2470	.2549	3.923	.9690	75.7°	1.3212
.2513	14.4°	.2487	.2568	3.895	.9686	75.6°	1.3195
.2531	14.5°	.2504	.2586	3.867	.9681	75.5°	1.3177
.2548	14.6°	.2521	.2605	3.839	.9677	75.4°	1.3160
.2566	14.7°	.2538	.2623	3.812	.9673	75.3°	1.3142
.2583	14.8°	.2554	.2642	3.785	.9668	75.2°	1.3125
.2601	14.9°	.2571	.2661	3.758	.9664	75.1°	1.3107
.2618	15.0°	.2588	.2679	3.732	.9659	75.0°	1.3090
		cos	cot	tan	sin	Degrees	Radians

TABLE 1

Radians	Degrees	sin	tan	cot	cos		
.2618	15.0°	.2588	.2679	3.732	.9659	75.0°	1.3090
.2635	15.1°	.2605	.2698	3.706	.9655	74.9°	1.3073
.2653	15.2°	.2622	.2717	3.681	.9650	74.8°	1.3055
.2670	15.3°	.2639	.2736	3.655	.9646	74.7°	1.3038
.2688	15.4°	.2656	.2754	3.630	.9641	74.6°	1.3020
.2705	15.5°	.2672	.2773	3.606	.9636	74.5°	1.3003
.2723	15.6°	.2689	.2792	3.582	.9632	74.4°	1.2985
.2740	15.7°	.2706	.2811	3.558	.9627	74.3°	1.2968
.2758	15.8°	.2723	.2830	3.534	.9622	74.2°	1.2950
.2775	15.9°	.2740	.2849	3.511	.9617	74.1°	1.2933
.2793	16.0°	.2756	.2867	3.487	.9613	74.0°	1.2915
.2810	16.1°	.2773	.2886	3.465	.9608	73.9°	1.2898
.2827	16.2°	.2790	.2905	3.442	.9603	73.8°	1.2881
.2845	16.3°	.2807	.2924	3.420	.9598	73.7°	1.2863
.2862	16.4°	.2823	.2943	3.398	.9593	73.6°	1.2846
.2880	16.5°	.2840	.2962	3.376	.9588	73.5°	1.2828
.2897	16.6°	.2857	.2981	3.354	.9583	73.4°	1.2811
.2915	16.7°	.2874	.3000	3.333	.9578	73.3°	1.2793
.2932	16.8°	.2890	.3019	3.312	.9573	73.2°	1.2776
.2950	16.9°	.2907	.3038	3.291	.9568	73.1°	1.2758
.2967	17.0°	.2924	.3057	3.271	.9563	73.0°	1.2741
.2985	17.1°	.2940	.3076	3.251	.9558	72.9°	1.2723
.3002	17.2°	.2957	.3096	3.230	.9553	72.8°	1.2706
.3019	17.3°	.2974	.3115	3.211	.9548	72.7°	1.2689
.3037	17.4°	.2990	.3134	3.191	.9542	72.6°	1.2671
.3054	17.5°	.3007	.3153	3.172	.9537	72.5°	1.2654
.3072	17.6°	.3024	.3172	3.152	.9532	72.4°	1.2636
.3089	17.7°	.3040	.3191	3.133	.9527	72.3°	1.2619
.3107	17.8°	.3057	.3211	3.115	.9521	72.2°	1.2601
.3124	17.9°	.3074	.3230	3.096	.9516	72.1°	1.2584
.3142	18.0°	.3090	.3249	3.078	.9511	72.0°	1.2566
.3159	18.1°	.3107	.3269	3.060	.9505	71.9°	1.2549
.3176	18.2°	.3123	.3288	3.042	.9500	71.8°	1.2531
.3194	18.3°	.3140	.3307	3.024	.9494	71.7°	1.2514
.3211	18.4°	.3156	.3327	3.006	.9489	71.6°	1.2497
.3229	18.5°	.3173	.3346	2.989	.9483	71.5°	1.2479
.3246	18.6°	.3190	.3365	2.971	.9478	71.4°	1.2462
.3264	18.7°	.3206	.3385	2.954	.9472	71.3°	1.2444
.3281	18.8°	.3223	.3404	2.937	.9466	71.2°	1.2427
.3299	18.9°	.3239	.3424	2.921	.9461	71.1°	1.2409
.3316	19.0°	.3256	.3443	2.904	.9455	71.0°	1.2392
.3334	19.1°	.3272	.3463	2.888	.9449	70.9°	1.2374
.3351	19.2°	.3289	.3482	2.872	.9444	70.8°	1.2357
.3368	19.3°	.3305	.3502	2.856	.9438	70.7°	1.2339
.3386	19.4°	.3322	.3522	2.840	.9432	70.6°	1.2322
.3403	19.5°	.3338	.3541	2.824	.9426	70.5°	1.2305
.3421	19.6°	.3355	.3561	2.808	.9421	70.4°	1.2287
.3438	19.7°	.3371	.3581	2.793	.9415	70.3°	1.2270
.3456	19.8°	.3387	.3600	2.778	.9409	70.2°	1.2252
.3473	19.9°	.3404	.3620	2.762	.9403	70.1°	1.2235
.3491	20.0°	.3420	.3640	2.747	.9397	70.0°	1.2217
		cos	cot	tan	sin	Degrees	Radians

TABLE 1

Radians	Degrees	sin	tan	cot	cos		
.3491	20.0°	.3420	.3640	2.747	.9397	70.0°	1.2217
.3508	20.1°	.3437	.3659	2.733	.9391	69.9°	1.2200
.3526	20.2°	.3453	.3679	2.718	.9385	69.8°	1.2182
.3543	20.3°	.3469	.3699	2.703	.9379	69.7°	1.2165
.3560	20.4°	.3486	.3719	2.689	.9373	69.6°	1.2147
.3578	20.5°	.3502	.3739	2.675	.9367	69.5°	1.2130
.3595	20.6°	.3518	.3759	2.660	.9361	69.4°	1.2113
.3613	20.7°	.3535	.3779	2.646	.9354	69.3°	1.2095
.3630	20.8°	.3551	.3799	2.633	.9348	69.2°	1.2078
.3648	20.9°	.3567	.3819	2.619	.9342	69.1°	1.2060
.3665	21.0°	.3584	.3839	2.605	.9336	69.0°	1.2043
.3683	21.1°	.3600	.3859	2.592	.9330	68.9°	1.2025
.3700	21.2°	.3616	.3879	2.578	.9323	68.8°	1.2008
.3718	21.3°	.3633	.3899	2.565	.9317	68.7°	1.1990
.3735	21.4°	.3649	.3919	2.552	.9311	68.6°	1.1973
.3752	21.5°	.3665	.3939	2.539	.9304	68.5°	1.1956
.3770	21.6°	.3681	.3959	2.526	.9298	68.4°	1.1938
.3787	21.7°	.3697	.3979	2.513	.9291	68.3°	1.1921
.3805	21.8°	.3714	.4000	2.500	.9285	68.2°	1.1903
.3822	21.9°	.3730	.4020	2.488	.9278	68.1°	1.1886
.3840	22.0°	.3746	.4040	2.475	.9272	68.0°	1.1868
.3857	22.1°	.3762	.4061	2.463	.9265	67.9°	1.1851
.3875	22.2°	.3778	.4081	2.450	.9259	67.8°	1.1833
.3892	22.3°	.3795	.4101	2.438	.9252	67.7°	1.1816
.3910	22.4°	.3811	.4122	2.426	.9245	67.6°	1.1798
.3927	22.5°	.3827	.4142	2.414	.9239	67.5°	1.1781
.3944	22.6°	.3843	.4163	2.402	.9232	67.4°	1.1764
.3962	22.7°	.3859	.4183	2.391	.9225	67.3°	1.1746
.3979	22.8°	.3875	.4204	2.379	.9219	67.2°	1.1729
.3997	22.9°	.3891	.4224	2.367	.9212	67.1°	1.1711
.4014	23.0°	.3907	.4245	2.356	.9205	67.0°	1.1694
.4032	23.1°	.3923	.4265	2.344	.9198	66.9°	1.1676
.4049	23.2°	.3939	.4286	2.333	.9191	66.8°	1.1659
.4067	23.3°	.3955	.4307	2.322	.9184	66.7°	1.1641
.4084	23.4°	.3971	.4327	2.311	.9178	66.6°	1.1624
.4102	23.5°	.3987	.4348	2.300	.9171	66.5°	1.1606
.4119	23.6°	.4003	.4369	2.289	.9164	66.4°	1.1589
.4136	23.7°	.4019	.4390	2.278	.9157	66.3°	1.1572
.4154	23.8°	.4035	.4411	2.267	.9150	66.2°	1.1554
.4171	23.9°	.4051	.4431	· 2.257	.9143	66.1°	1.1537
.4189	24.0°	.4067	.4452	2.246	.9135	66.0°	1.1519
.4206	24.1°	.4083	.4473	2.236	.9128	65.9°	1.1502
.4224	24.2°	.4099	.4494	2.225	.9121	65.8°	1.1484
.4241	24.3°	.4115	.4515	2.215	.9114	65.7°	1.1467
.4259	24.4°	.4131	.4536	2.204	.9107	65.6°	1.1449
.4276	24.5°	.4147	.4557	2.194	.9100	65.5°	1.1432
.4294	24.6°	.4163	.4578	2.184	.9092	65.4°	1.1414
.4311	24.7°	.4179	.4599	2.174	.9085	65.3°	1.1397
.4328	24.8°	.4195	.4621	2.164	.9078	65.2°	1.1380
.4346	24.9°	.4210	.4642	2.154	.9070	65.1°	1.1362
.4363	25.0°	.4226	.4663	2.145	.9063	65.0°	1.1345
		cos	cot	tan	sin	Degrees	Radians

TABLE 1

Radians	Degrees	sin	tan	cot	cos		
.4363	25.0°	.4226	.4663	2.145	.9063	65.0°	1.1345
.4381	25.1°	.4242	.4684	2.135	.9056	64.9°	1.1327
.4398	25.2°	.4258	.4706	2.125	.9048	64.8°	1.1310
.4416	25.3°	.4274	.4727	2.116	.9041	64.7°	1.1292
.4433	25.4°	.4289	.4748	2.106	.9033	64.6°	1.1275
.4451	25.5°	.4305	.4770	2.097	.9026	64.5°	1.1257
.4468	25.6°	.4321	.4791	2.087	.9018	64.4°	1.1240
.4485	25.7°	.4337	.4813	2.078	.9011	64.3°	1.1222
.4503	25.8°	.4352	.4834	2.069	.9003	64.2°	1.1205
.4520	25.9°	.4368	.4856	2.059	.8996	64.1°	1.1188
.4538	26.0°	.4384	.4877	2.050	.8988	64.0°	1.1170
.4555	26.1°	.4399	.4899	2.041	.8980	63.9°	1.1153
.4573	26.2°	.4415	.4921	2.032	.8973	63.8°	1.1135
.4590	26.3°	.4431	.4942	2.023	.8965	63.7°	1.1118
.4608	26.4°	.4446	.4964	2.014	.8957	63.6°	1.1100
.4625	26.5°	.4462	.4986	2.006	.8949	63.5°	1.1083
.4643	26.6°	.4478	.5008	1.997	.8942	63.4°	1.1065
.4660	26.7°	.4493	.5029	1.988	.8934	63.3°	1.1048
.4677	26.8°	.4509	.5051	1.980	.8926	63.2°	1.1030
.4695	26.9°	.4524	.5073	1.971	.8918	63.1°	1.1013
.4712	27.0°	.4540	.5095	1.963	.8910	63.0°	1.0996
.4730	27.1°	.4555	.5117	1.954	.8902	62.9°	1.0978
.4747	27.2°	.4571	.5139	1.946	.8894	62.8°	1.0961
.4765	27.3°	.4586	.5161	1.937	.8886	62.7°	1.0943
.4782	27.4°	.4602	.5184	1.929	.8878	62.6°	1.0926
.4800	27.5°	.4617	.5206	1.921	.8870	62.5°	1.0908
.4817	27.6°	.4633	.5228	1.913	.8862	62.4°	1.0891
.4835	27.7°	.4648	.5250	1.905	.8854	62.3°	1.0873
.4852	27.8°	.4664	.5272	1.897	.8846	62.2°	1.0856
.4869	27.9°	.4679	.5295	1.889	.8838	62.1°	1.0838
.4887	28.0°	.4695	.5317	1.881	.8829	62.0°	1.0821
.4904	28.1°	.4710	.5340	1.873	.8821	61.9°	1.0804
.4922	28.2°	.4726	.5362	1.865	.8813	61.8°	1.0786
.4939	28.3°	.4741	.5384	1.857	.8805	61.7°	1.0769
.4957	28.4°	.4756	.5407	1.849	.8796	61.6°	1.0751
.4974	28.5°	.4772	.5430	1.842	.8788	61.5°	1.0734
.4992	28.6°	.4787	.5452	1.834	.8780	61.4°	1.0716
.5009	28.7°	.4802	.5475	1.827	.8771	61.3°	1.0699
.5027	28.8°	.4818	.5498	1.819	.8763	61.2°	1.0681
.5044	28.9°	.4833	.5520	1.811	.8755	61.1°	1.0664
.5061	29.0°	.4848	.5543	1.804	.8746	61.0°	1.0647
.5079	29.1°	.4863	.5566	1.797	.8738	60.9°	1.0629
.5096	29.2°	.4879	.5589	1.789	.8729	60.8°	1.0612
.5114	29.3°	.4894	.5612	1.782	.8721	60.7°	1.0594
.5131	29.4°	.4909	.5635	1.775	.8712	60.6°	1.0577
.5149	29.5°	.4924	.5658	1.767	.8704	60.5°	1.0559
.5166	29.6°	.4939	.5681	1.760	.8695	60.4°	1.0542
.5184	29.7°	.4955	.5704	1.753	.8686	60.3°	1.0524
.5201	29.8°	.4970	.5727	1.746	.8678	60.2°	1.0507
.5219	29.9°	.4985	.5750	1.739	.8669	60.1°	1.0489
.5236	30.0°	.5000	.5774	1.732	.8660	60.0°	1.0472
		cos	cot	tan	sin	Degrees	Radians

TABLE 1

Radians	Degrees	sin	tan	cot	cos		
.5236	30.0°	.5000	.5774	1.732	.8660	60.0°	1.0472
.5253	30.1°	.5015	.5797	1.725	.8652	59.9°	1.0455
.5271	30.2°	.5030	.5820	1.718	.8643	59.8°	1.0437
.5288	30.3°	.5045	.5844	1.711	.8634	59.7°	1.0420
.5306	30.4°	.5060	.5867	1.704	.8625	59.6°	1.0402
.5323	30.5°	.5075	.5890	1.698	.8616	59.5°	1.0385
.5341	30.6°	.5090	.5914	1.691	.8607	59.4°	1.0367
.5358	30.7°	.5105	.5938	1.684	.8599	59.3°	1.0350
.5376	30.8°	.5120	.5961	1.678	.8590	59.2°	1.0332
.5393	30.9°	.5135	.5985	1.671	.8581	59.1°	1.0315
.5411	31.0°	.5150	.6009	1.664	.8572	59.0°	1.0297
.5428	31.1°	.5165	.6032	1.658	.8563	58.9°	1.0280
.5445	31.2°	.5180	.6056	1.651	.8554	58.8°	1.0263
.5463	31.3°	.5195	.6080	1.645	.8545	58.7°	1.0245
.5480	31.4°	.5210	.6104	1.638	.8536	58.6°	1.0228
.5498	31.5°	.5225	.6128	1.632	.8526	58.5°	1.0210
.5515	31.6°	.5240	.6152	1.625	.8517	58.4°	1.0193
.5533	31.7°	.5255	.6176	1.619	.8508	58.3°	1.0175
.5550	31.8°	.5270	.6200	1.613	.8499	58.2°	1.0158
.5568	31.9°	.5284	.6224	1.607	.8490	58.1°	1.0140
.5585	32.0°	.5299	.6249	1.600	.8480	58.0°	1.0123
.5603	32.1°	.5314	.6273	1.594	.8471	57.9°	1.0105
.5620	32.2°	.5329	.6297	1.588	.8462	57.8°	1.0088
.5637	32.3°	.5344	.6322	1.582	.8453	57.7°	1.0071
.5655	32.4°	.5358	.6346	1.576	.8443	57.6°	1.0053
.5672	32.5°	.5373	.6371	1.570	.8434	57.5°	1.0036
.5690	32.6°	.5388	.6395	1.564	.8425	57.4°	1.0018
.5707	32.7°	.5402	.6420	1.558	.8415	57.3°	1.0001
.5725	32.8°	.5417	.6445	1.552	.8406	57.2°	.9983
.5742	32.9°	.5432	.6469	1.546	.8396	57.1°	.9966
.5760	33.0°	.5446	.6494	1.540	.8387	57.0°	.9948
.5777	33.1°	.5461	.6519	1.534	.8377	56.9°	.9931
.5794	33.2°	.5476	.6544	1.528	.8368	56.8°	.9913
.5812	33.3°	.5490	.6569	1.522	.8358	56.7°	.9896
.5829	33.4°	.5505	.6594	1.517	.8348	56.6°	.9879
.5847	33.5°	.5519	.6619	1.511	.8339	56.5°	.9861
.5864	33.6°	.5534	.6644	1.505	.8329	56.4°	.9844
.5882	33.7°	.5548	.6669	1.499	.8320	56.3°	.9826
.5899	33.8°	.5563	.6694	1.494	.8310	56.2°	.9809
.5917	33.9°	.5577	.6720	.1488	.8300	56.1°	.9791
.5934	34.0°	.5592	.6745	1.483	.8290	56.0°	.9774
.5952	34.1°	.5606	.6771	1.477	.8281	55.9°	.9756
.5969	34.2°	.5621	.6796	1.471	.8271	55.8°	.9739
.5986	34.3°	.5635	.6822	1.466	.8261	55.7°	.9721
.6004	34.4°	.5650	.6847	1.460	.8251	55.6°	.9704
.6021	34.5°	.5664	.6873	1.455	.8241	55.5°	.9687
.6039	34.6°	.5678	.6899	1.450	.8231	55.4°	.9669
.6056	34.7°	.5693	.6924	1.444	.8221	55.3°	.9652
.6074	34.8°	.5707	.6950	1.439	.8211	55.2°	.9634
.6091	34.9°	.5721	.6976	1.433	.8202	55.1°	.9617
.6109	35.0°	.5736	.7002	1.428	.8192	55.0°	.9599
		cos	cot	tan	sin	Degrees	Radians

TABLE 1

Radians	Degrees	sin	tan	cot	cos		
.6109	35.0°	.5736	.7002	1.428	.8192	55.0°	.9599
.6126	35.1°	.5750	.7028	1.423	.8181	54.9°	.9582
.6144	35.2°	.5764	.7054	1.418	.8171	54.8°	.9564
.6161	35.3°	.5779	.7080	1.412	.8161	54.7°	.9547
.6178	35.4°	.5793	.7107	1.407	.8151	54.6°	.9530
.6196	35.5°	.5807	.7133	1.402	.8141	54.5°	.9512
.6213	35.6°	.5821	.7159	1.397	.8131	54.4°	.9495
.6231	35.7°	.5835	.7186	1.392	.8121	54.3°	.9477
.6248	35.8°	.5850	.7212	1.387	.8111	54.2°	.9460
.6266	35.9°	.5864	.7239	1.381	.8100	54.1°	.9442
.6283	36.0°	.5878	.7265	1.376	.8090	54.0°	.9425
.6301	36.1°	.5892	.7292	1.371	.8080	53.9°	.9407
.6318	36.2°	.5906	.7319	1.366	.8070	53.8°	.9390
.6336	36.3°	.5920	.7346	1.361	.8059	53.7°	.9372
.6353	36.4°	.5934	.7373	1.356	.8049	53.6°	.9355
.6370	36.5°	.5948	.7400	1.351	.8039	53.5°	.9338
.6388	36.6°	.5962	.7427	1.347	.8028	53.4°	.9320
.6405	36.7°	.5976	.7454	1.342	.8018	53.3°	.9303
.6423	36.8°	.5990	.7481	1.337	.8007	53.2°	.9285
.6440	36.9°	.6004	.7508	1.332	.7997	53.1°	.9268
.6458	37.0°	.6018	.7536	1.327	.7986	53.0°	.9250
.6475	37.1°	.6032	.7563	1.322	.7976	52.9°	.9233
.6493	37.2°	.6046	.7590	1.317	.7965	52.8°	.9215
.6510	37.3°	.6060	.7618	1.313	.7955	52.7°	.9198
.6528	37.4°	.6074	.7646	1.308	.7944	52.6°	.9180
.6545	37.5°	.6088	.7673	1.303	.7934	52.5°	.9163
.6562	37.6°	.6101	.7701	1.299	.7923	52.4°	.9146
.6580	37.7°	.6115	.7729	1.294	.7912	52.3°	.9128
.6597	37.8°	.6129	.7757	1.289	.7902	52.2°	.9111
.6615	37.9°	.6143	.7785	1.285	.7891	52.1°	.9093
.6632	38.0°	.6157	.7813	1.280	.7880	52.0°	.9076
.6650	38.1°	.6170	.7841	1.275	.7869	51.9°	.9058
.6667	38.2°	.6184	.7869	1.271	.7859	51.8°	.9041
.6685	38.3°	.6198	.7898	1.266	.7848	51.7°	.9023
.6702	38.4°	.6211	.7926	1.262	.7837	51.6°	.9006
.6720	38.5°	.6225	.7954	1.257	.7826	51.5°	.8988
.6737	38.6°	.6239	.7983	1.253	.7815	51.4°	.8971
.6754	38.7°	.6252	.8012	1.248	.7804	51.3°	.8954
.6772	38.8°	.6266	.8040	1.244	.7793	51.2°	.8936
.6789	38.9°	.6280	.8069	1.239	.7782	51.1°	.8919
.6807	39.0°	.6293	.8098	1.235	.7771	51.0°	.8901
.6824	39.1°	.6307	.8127	1.231	.7760	50.9°	.8884
.6842	39.2°	.6320	.8156	1.226	.7749	50.8°	.8866
.6859	39.3°	.6334	.8185	1.222	.7738	50.7°.	.8849
.6877	39.4°	.6347	.8214	1.217	.7727	50.6°	.8831
.6894	39.5°	.6361	.8243	1.213	.7716	50.5°	.8814
.6912	39.6°	.6374	.8273	1.209	.7705	50.4°	.8796
.6929	39.7°	.6388	.8302	1.205	.7694	50.3°	.8779
.6946	39.8°	.6401	.8332	1.200	.7683	50.2°	.8762
.6964	39.9°	.6414	.8361	1.196	.7672	50.1°	.8744
.6981	40.0°	.6428	.8391	1.192	.7660	50.0°	.8727

| | | cos | cot | tan | sin | Degrees | Radians |

TABLE 1

Radians	Degrees	sin	tan	cot	cos		
.6981	40.0°	.6428	.8391	1.192	.7660	50.0°	.8727
.6999	40.1°	.6441	.8421	1.188	.7649	49.9°	.8709
.7016	40.2°	.6455	.8451	1.183	.7638	49.8°	.8692
.7034	40.3°	.6468	.8481	1.179	.7627	49.7°	.8674
.7051	40.4°	.6481	.8511	1.175	.7615	49.6°	.8657
.7069	40.5°	.6494	.8541	1.171	.7604	49.5°	.8639
.7086	40.6°	.6508	.8571	1.167	.7593	49.4°	.8622
.7103	40.7°	.6521	.8601	1.163	.7581	49.3°	.8604
.7121	40.8°	.6534	.8632	1.159	.7570	49.2°	.8587
.7138	40.9°	.6547	.8662	1.154	.7559	49.1°	.8570
.7156	41.0°	.6561	.8693	1.150	.7547	49.0°	.8552
.7173	41.1°	.6574	.8724	1.146	.7536	48.9°	.8535
.7191	41.2°	.6587	.8754	1.142	.7524	48.8°	.8517
.7208	41.3°	.6600	.8785	1.138	.7513	48.7°	.8500
.7226	41.4°	.6613	.8816	1.134	.7501	48.6°	.8482
.7243	41.5°	.6626	.8847	1.130	.7490	48.5°	.8465
.7261	41.6°	.6639	.8878	1.126	.7478	48.4°	.8447
.7278	41.7°	.6652	.8910	1.122	.7466	48.3°	.8430
.7295	41.8°	.6665	.8941	1.118	.7455	48.2°	.8412
.7313	41.9°	.6678	.8972	1.115	.7443	48.1°	.8395
.7330	42.0°	.6691	.9004	1.111	.7431	48.0°	.8378
.7348	42.1°	.6704	.9036	1.107	.7420	47.9°	.8360
.7365	42.2°	.6717	.9067	1.103	.7408	47.8°	.8343
.7383	42.3°	.6730	.9099	1.099	.7396	47.7°	.8325
.7400	42.4°	.6743	.9131	1.095	.7385	47.6°	.8308
.7418	42.5°	.6756	.9163	1.091	.7373	47.5°	.8290
.7435	42.6°	.6769	.9195	1.087	.7361	47.4°	.8273
.7453	42.7°	.6782	.9228	1.084	.7349	47.3°	.8255
.7470	42.8°	.6794	.9260	1.080	.7337	47.2°	.8238
.7487	42.9°	.6807	.9293	1.076	.7325	47.1°	.8221
.7505	43.0°	.6820	.9325	1.072	.7314	47.0°	.8203
.7522	43.1°	.6833	.9358	1.069	.7302	46.9°	.8186
.7540	43.2°	.6845	.9391	1.065	.7290	46.8°	.8168
.7557	43.3°	.6858	.9424	1.061	.7278	46.7°	.8151
.7575	43.4°	.6871	.9457	1.057	.7266	46.6°	.8133
.7592	43.5°	.6884	.9490	1.054	.7254	46.5°	.8116
.7610	43.6°	.6896	.9523	1.050	.7242	46.4°	.8098
.7627	43.7°	.6909	.9556	1.046	.7230	46.3°	.8081
.7645	43.8°	.6921	.9590	1.043	.7218	46.2°	.8063
.7662	43.9°	.6934	.9623	1.039	.7206	46.1°	.8046
.7679	44.0°	.6947	.9657	1.036	.7193	46.0°	.8029
.7697	44.1°	.6959	.9691	1.032	.7181	45.9°	.8011
.7714	44.2°	.6972	.9725	1.028	.7169	45.8°	.7994
.7732	44.3°	.6984	.9759	1.025	.7157	45.7°	.7976
.7749	44.4°	.6997	.9793	1.021	.7145	45.6°	.7959
.7767	44.5°	.7009	.9827	1.018	.7133	45.5°	.7941
.7784	44.6°	.7022	.9861	1.014	.7120	45.4°	.7924
.7802	44.7°	.7034	.9896	1.011	.7108	45.3°	.7906
.7819	44.8°	.7046	.9930	1.007	.7096	45.2°	.7889
.7837	44.9°	.7059	.9965	1.003	.7083	45.1°	.7871
.7854	45.0°	.7071	1.0000	1.000	.7071	45.0°	.7854
		cos	cot	tan	sin	Degrees	Radians

TABLE 2   MANTISSAS OF COMMON LOGARITHMS

N.	0	1	2	3	4	5	6	7	8	9
10	0000	0043	0086	0128	0170	0212	0253	0294	0334	0374
11	0414	0453	0492	0531	0569	0607	0645	0682	0719	0755
12	0792	0828	0864	0899	0934	0969	1004	1038	1072	1106
13	1139	1173	1206	1239	1271	1303	1335	1367	1399	1430
14	1461	1492	1523	1553	1584	1614	1644	1673	1703	1732
15	1761	1790	1818	1847	1875	1903	1931	1959	1987	2014
16	2041	2068	2095	2122	2148	2175	2201	2227	2253	2279
17	2304	2330	2355	2380	2405	2430	2455	2480	2504	2529
18	2553	2577	2601	2625	2648	2672	2695	2718	2742	2765
19	2788	2810	2833	2856	2878	2900	2923	2945	2967	2989
20	3010	3032	3054	3075	3096	3118	3139	3160	3181	3201
21	3222	3243	3263	3284	3304	3324	3345	3365	3385	3404
22	3424	3444	3464	3483	3502	3522	3541	3560	3579	3598
23	3617	3636	3655	3674	3692	3711	3729	3747	3766	3784
24	3802	3820	3838	3856	3874	3892	3909	3927	3945	3962
25	3979	3997	4014	4031	4048	4065	4082	4099	4116	4133
26	4150	4166	4183	4200	4216	4232	4249	4265	4281	4298
27	4314	4330	4346	4362	4378	4393	4409	4425	4440	4456
28	4472	4487	4502	4518	4533	4548	4564	4579	4594	4609
29	4624	4639	4654	4669	4683	4698	4713	4728	4742	4757
30	4771	4786	4800	4814	4829	4843	4857	4871	4886	4900
31	4914	4928	4942	4955	4969	4983	4997	5011	5024	5038
32	5051	5065	5079	5092	5105	5119	5132	5145	5159	5172
33	5185	5198	5211	5224	5237	5250	5263	5276	5289	5302
34	5315	5328	5340	5353	5366	5378	5391	5403	5416	5428
35	5441	5453	5465	5478	5490	5502	5514	5527	5539	5551
36	5563	5575	5587	5599	5611	5623	5635	5647	5658	5670
37	5682	5694	5705	5717	5729	5740	5752	5763	5775	5786
38	5798	5809	5821	5832	5843	5855	5866	5877	5888	5899
39	5911	5922	5933	5944	5955	5966	5977	5988	5999	6010
40	6021	6031	6042	6053	6064	6075	6085	6096	6107	6117
41	6128	6138	6149	6160	6170	6180	6191	6201	6212	6222
42	6232	6243	6253	6263	6274	6284	6294	6304	6314	6325
43	6335	6345	6355	6365	6375	6385	6395	6405	6415	6425
44	6435	6444	6454	6464	6474	6484	6493	6503	6513	6522
45	6532	6542	6551	6561	6571	6580	6590	6599	6609	6618
46	6628	6637	6646	6656	6665	6675	6684	6693	6702	6712
47	6721	6730	6739	6749	6758	6767	6776	6785	6794	6803
48	6812	6821	6830	6839	6848	6857	6866	6875	6884	6893
49	6902	6911	6920	6928	6937	6946	6955	6964	6972	6981
50	6990	6998	7007	7016	7024	7033	7042	7050	7059	7067
51	7076	7084	7093	7101	7110	7118	7126	7135	7143	7152
52	7160	7168	7177	7185	7193	7202	7210	7218	7226	7235
53	7243	7251	7259	7267	7275	7284	7292	7300	7308	7316
54	7324	7332	7340	7348	7356	7364	7372	7380	7388	7396
N.	0	1	2	3	4	5	6	7	8	9

TABLE 2 MANTISSAS OF COMMON LOGARITHMS

N.	0	1	2	3	4	5	6	7	8	9
55	7404	7412	7419	7427	7435	7443	7451	7459	7466	7474
56	7482	7490	7497	7505	7513	7520	7528	7536	7543	7551
57	7559	7566	7574	7582	7589	7597	7604	7612	7619	7627
58	7634	7642	7649	7657	7664	7672	7679	7686	7694	7701
59	7709	7716	7723	7731	7738	7745	7752	7760	7767	7774
60	7782	7789	7796	7803	7810	7818	7825	7832	7839	7846
61	7853	7860	7868	7875	7882	7889	7896	7903	7910	7917
62	7924	7931	7938	7945	7952	7959	7966	7973	7980	7987
63	7993	8000	8007	8014	8021	8028	8035	8041	8048	8055
64	8062	8069	8075	8082	8089	8096	8102	8109	8116	8122
65	8129	8136	8142	8149	8156	8162	8169	8176	8182	8189
66	8195	8202	8209	8215	8222	8228	8235	8241	8248	8254
67	8261	8267	8274	8280	8287	8293	8299	8306	8312	8319
68	8325	8331	8338	8344	8351	8357	8363	8370	8376	8382
69	8388	8395	8401	8407	8414	8420	8426	8432	8439	8445
70	8451	8457	8463	8470	8476	8482	8488	8494	8500	8506
71	8513	8519	8525	8531	8537	8543	8549	8555	8561	8567
72	8573	8579	8585	8591	8597	8603	8609	8615	8621	8627
73	8633	8639	8645	8651	8657	8663	8669	8675	8681	8686
74	8692	8698	8704	8710	8716	8722	8727	8733	8739	8745
75	8751	8756	8762	8768	8774	8779	8785	8791	8797	8802
76	8808	8814	8820	8825	8831	8837	8842	8848	8854	8859
77	8865	8871	8876	8882	8887	8893	8899	8904	8910	8915
78	8921	8927	8932	8938	8943	8949	8954	8960	8965	8971
79	8976	8982	8987	8993	8998	9004	9009	9015	9020	9025
80	9031	9036	9042	9047	9053	9058	9063	9069	9074	9079
81	9085	9090	9096	9101	9106	9112	9117	9122	9128	9133
82	9138	9143	9149	9154	9159	9165	9170	9175	9180	9186
83	9191	9196	9201	9206	9212	9217	9222	9227	9232	9238
84	9243	9248	9253	9258	9263	9269	9274	9279	9284	9289
85	9294	9299	9304	9309	9315	9320	9325	9330	9335	9340
86	9345	9350	9355	9360	9365	9370	9375	9380	9385	9390
87	9395	9400	9405	9410	9415	9420	9425	9430	9435	9440
88	9445	9450	9455	9460	9465	9469	9474	9479	9484	9489
89	9494	9499	9504	9509	9513	9518	9523	9528	9533	9538
90	9542	9547	9552	9557	9562	9566	9571	9576	9581	9586
91	9590	9595	9600	9605	9609	9614	9619	9624	9628	9633
92	9638	9643	9647	9652	9657	9661	9666	9671	9675	9680
93	9685	9689	9694	9699	9703	9708	9713	9717	9722	9727
94	9731	9736	9741	9745	9750	9754	9759	9763	9768	9773
95	9777	9782	9786	9791	9795	9800	9805	9809	9814	9818
96	9823	9827	9832	9836	9841	9845	9850	9854	9859	9863
97	9868	9872	9877	9881	9886	9890	9894	9899	9903	9908
98	9912	9917	9921	9926	9930	9934	9939	9943	9948	9952
99	9956	9961	9965	9969	9974	9978	9983	9987	9991	9996
N.	0	1	2	3	4	5	6	7	8	9

TABLE 3 (Subtract 10 from Each Entry)

Degrees	L sin	L tan	L cot	L cos	
0.0°	—	—	—	10.0000	90.0°
0.1°	7.2419	7.2419	12.7581	10.0000	89.9°
0.2°	7.5429	7.5429	12.4571	10.0000	89.8°
0.3°	7.7190	7.7190	12.2810	10.0000	89.7°
0.4°	7.8439	7.8439	12.1561	10.0000	89.6°
0.5°	7.9408	7.9409	12.0591	10.0000	89.5°
0.6°	8.0200	8.0200	11.9800	10.0000	89.4°
0.7°	8.0870	8.0870	11.9130	10.0000	89.3°
0.8°	8.1450	8.1450	11.8550	10.0000	89.2°
0.9°	8.1961	8.1962	11.8038	9.9999	89.1°
1.0°	8.2419	8.2419	11.7581	9.9999	89.0°
1.1°	8.2832	8.2833	11.7167	9.9999	88.9°
1.2°	8.3210	8.3211	11.6789	9.9999	88.8°
1.3°	8.3558	8.3559	11.6441	9.9999	88.7°
1.4°	8.3880	8.3881	11.6119	9.9999	88.6°
1.5°	8.4179	8.4181	11.5819	9.9999	88.5°
1.6°	8.4459	8.4461	11.5539	9.9998	88.4°
1.7°	8.4723	8.4725	11.5275	9.9998	88.3°
1.8°	8.4971	8.4973	11.5027	9.9998	88.2°
1.9°	8.5206	8.5208	11.4792	9.9998	88.1°
2.0°	8.5428	8.5431	11.4569	9.9997	88.0°
2.1°	8.5640	8.5643	11.4357	9.9997	87.9°
2.2°	8.5842	8.5845	11.4155	9.9997	87.8°
2.3°	8.6035	8.6038	11.3962	9.9997	87.7°
2.4°	8.6220	8.6223	11.3777	9.9996	87.6°
2.5°	8.6397	8.6401	11.3599	9.9996	87.5°
2.6°	8.6567	8.6571	11.3429	9.9996	87.4°
2.7°	8.6731	8.6736	11.3264	9.9995	87.3°
2.8°	8.6889	8.6894	11.3106	9.9995	87.2°
2.9°	8.7041	8.7046	11.2954	9.9994	87.1°
3.0°	8.7188	8.7194	11.2806	9.9994	87.0°
3.1°	8.7330	8.7337	11.2663	9.9994	86.9°
3.2°	8.7468	8.7475	11.2525	9.9993	86.8°
3.3°	8.7602	8.7609	11.2391	9.9993	86.7°
3.4°	8.7731	8.7739	11.2261	9.9992	86.6°
3.5°	8.7857	8.7865	11.2135	9.9992	86.5°
3.6°	8.7979	8.7988	11.2012	9.9991	86.4°
3.7°	8.8098	8.8107	11.1893	9.9991	86.3°
3.8°	8.8213	8.8223	11.1777	9.9990	86.2°
3.9°	8.8326	8.8336	11.1664	9.9990	86.1°
4.0°	8.8436	8.8446	11.1554	9.9989	86.0°
4.1°	8.8543	8.8554	11.1446	9.9989	85.9°
4.2°	8.8647	8.8659	11.1341	9.9988	85.8°
4.3°	8.8749	8.8762	11.1238	9.9988	85.7°
4.4°	8.8849	8.8862	11.1138	9.9987	85.6°
4.5°	8.8946	8.8960	11.1040	9.9987	85.5°
4.6°	8.9042	8.9056	11.0944	9.9986	85.4°
4.7°	8.9135	8.9150	11.0850	9.9985	85.3°
4.8°	8.9226	8.9241	11.0759	9.9985	85.2°
4.9°	8.9315	8.9331	11.0669	9.9984	85.1°
5.0°	8.9403	8.9420	11.0580	9.9983	85.0°
	L cos	L cot	L tan	L sin	Degrees

TABLE 3   (Subtract 10 from Each Entry)

Degrees	L sin	L tan	L cot	L cos	
5.0°	8.9403	8.9420	11.0580	9.9983	85.0°
5.1°	8.9489	8.9506	11.0494	9.9983	84.9°
5.2°	8.9573	8.9591	11.0409	9.9982	84.8°
5.3°	8.9655	8.9674	11.0326	9.9981	84.7°
5.4°	8.9736	8.9756	11.0244	9.9981	84.6°
5.5°	8.9816	8.9836	11.0164	9.9980	84.5°
5.6°	8.9894	8.9915	11.0085	9.9979	84.4°
5.7°	8.9970	8.9992	11.0008	9.9978	84.3°
5.8°	9.0046	9.0068	10.9932	9.9978	84.2°
5.9°	9.0120	9.0143	10.9857	9.9977	84.1°
6.0°	9.0192	9.0216	10.9784	9.9976	84.0°
6.1°	9.0264	9.0289	10.9711	9.9975	83.9°
6.2°	9.0334	9.0360	10.9640	9.9975	83.8°
6.3°	9.0403	9.0430	10.9570	9.9974	83.7°
6.4°	9.0472	9.0499	10.9501	9.9973	83.6°
6.5°	9.0539	9.0567	10.9433	9.9972	83.5°
6.6°	9.0605	9.0633	10.9367	9.9971	83.4°
6.7°	9.0670	9.0699	10.9301	9.9970	83.3°
6.8°	9.0734	9.0764	10.9236	9.9969	83.2°
6.9°	9.0797	9.0828	10.9172	9.9968	83.1°
7.0°	9.0859	9.0891	10.9109	9.9968	83.0°
7.1°	9.0920	9.0954	10.9046	9.9967	82.9°
7.2°	9.0981	9.1015	10.8985	9.9966	82.8°
7.3°	9.1040	9.1076	10.8924	9.9965	82.7°
7.4°	9.1099	9.1135	10.8865	9.9964	82.6°
7.5°	9.1157	9.1194	10.8806	9.9963	82.5°
7.6°	9.1214	9.1252	10.8748	9.9962	82.4°
7.7°	9.1271	9.1310	10.8690	9.9961	82.3°
7.8°	9.1326	9.1367	10.8633	9.9960	82.2°
7.9°	9.1381	9.1423	10.8577	9.9959	82.1°
8.0°	9.1436	9.1478	10.8522	9.9958	82.0°
8.1°	9.1489	9.1533	10.8467	9.9956	81.9°
8.2°	9.1542	9.1587	10.8413	9.9955	81.8°
8.3°	9.1594	9.1640	10.8360	9.9954	81.7°
8.4°	9.1646	9.1693	10.8307	9.9953	81.6°
8.5°	9.1697	9.1745	10.8255	9.9952	81.5°
8.6°	9.1747	9.1797	10.8203	9.9951	81.4°
8.7°	9.1797	9.1848	10.8152	9.9950	81.3°
8.8°	9.1847	9.1898	10.8102	9.9949	81.2°
8.9°	9.1895	9.1948	10.8052	9.9947	81.1°
9.0°	9.1943	9.1997	10.8003	9.9946	81.0°
9.1°	9.1991	9.2046	10.7954	9.9945	80.9°
9.2°	9.2038	9.2094	10.7906	9.9944	80.8°
9.3°	9.2085	9.2142	10.7858	9.9943	80.7°
9.4°	9.2131	9.2189	10.7811	9.9941	80.6°
9.5°	9.2176	9.2236	10.7764	9.9940	80.5°
9.6°	9.2221	9.2282	10.7718	9.9939	80.4°
9.7°	9.2266	9.2328	10.7672	9.9937	80.3°
9.8°	9.2310	9.2374	10.7626	9.9936	80.2°
9.9°	9.2353	9.2419	10.7581	9.9935	80.1°
10.0°	9.2397	9.2463	10.7537	9.9934	80.0°
	L cos	L cot	L tan	L sin	Degrees

TABLE 3 (Subtract 10 from Each Entry)

Degrees	L sin	L tan	L cot	L cos	
10.0°	9.2397	9.2463	10.7537	9.9934	80.0°
10.1°	9.2439	9.2507	10.7493	9.9932	79.9°
10.2°	9.2482	9.2551	10.7449	9.9931	79.8°
10.3°	9.2524	9.2594	10.7406	9.9929	79.7°
10.4°	9.2565	9.2637	10.7363	9.9928	79.6°
10.5°	9.2606	9.2680	10.7320	9.9927	79.5°
10.6°	9.2647	9.2722	10.7278	9.9925	79.4°
10.7°	9.2687	9.2764	10.7236	9.9924	79.3°
10.8°	9.2727	9.2805	10.7195	9.9922	79.2°
10.9°	9.2767	9.2846	10.7154	9.9921	79.1°
11.0°	9.2806	9.2887	10.7113	9.9919	79.0°
11.1°	9.2845	9.2927	10.7073	9.9918	78.9°
11.2°	9.2883	9.2967	10.7033	9.9916	78.8°
11.3°	9.2921	9.3006	10.6994	9.9915	78.7°
11.4°	9.2959	9.3046	10.6954	9.9913	78.6°
11.5°	9.2997	9.3085	10.6915	9.9912	78.5°
11.6°	9.3034	9.3123	10.6877	9.9910	78.4°
11.7°	9.3070	9.3162	10.6838	9.9909	78.3°
11.8°	9.3107	9.3200	10.6800	9.9907	78.2°
11.9°	9.3143	9.3237	10.6763	9.9906	78.1°
12.0°	9.3179	9.3275	10.6725	9.9904	78.0°
12.1°	9.3214	9.3312	10.6688	9.9902	77.9°
12.2°	9.3250	9.3349	10.6651	9.9901	77.8°
12.3°	9.3284	9.3385	10.6615	9.9899	77.7°
12.4°	9.3319	9.3422	10.6578	9.9897	77.6°
12.5°	9.3353	9.3458	10.6542	9.9896	77.5°
12.6°	9.3387	9.3493	10.6507	9.9894	77.4°
12.7°	9.3421	9.3529	10.6471	9.9892	77.3°
12.8°	9.3455	9.3564	10.6436	9.9891	77.2°
12.9°	9.3488	9.3599	10.6401	9.9889	77.1°
13.0°	9.3521	9.3634	10.6366	9.9887	77.0°
13.1°	9.3554	9.3668	10.6332	9.9885	76.9°
13.2°	9.3586	9.3702	10.6298	9.9884	76.8°
13.3°	9.3618	9.3736	10.6264	9.9882	76.7°
13.4°	9.3650	9.3770	10.6230	9.9880	76.6°
13.5°	9.3682	9.3804	10.6196	9.9878	76.5°
13.6°	9.3713	9.3837	10.6163	9.9876	76.4°
13.7°	9.3745	9.3870	10.6130	9.9875	76.3°
13.8°	9.3775	9.3903	10.6097	9.9873	76.2°
13.9°	9.3806	9.3935	10.6065	9.9871	76.1°
14.0°	9.3837	9.3968	10.6032	9.9869	76.0°
14.1°	9.3867	9.4000	10.6000	9.9867	75.9°
14.2°	9.3897	9.4032	10.5968	9.9865	75.8°
14.3°	9.3927	9.4064	10.5936	9.9863	75.7°
14.4°	9.3957	9.4095	10.5905	9.9861	75.6°
14.5°	9.3986	9.4127	10.5873	9.9859	75.5°
14.6°	9.4015	9.4158	10.5842	9.9857	75.4°
14.7°	9.4044	9.4189	10.5811	9.9855	75.3°
14.8°	9.4073	9.4220	10.5780	9.9853	75.2°
14.9°	9.4102	9.4250	10.5750	9.9851	75.1°
15.0°	9.4130	9.4281	10.5719	9.9849	75.0°
	L cos	L cot	L tan	L sin	Degrees

TABLE 3 (Subtract 10 from Each Entry)

Degrees	L sin	L tan	L cot	L cos	
15.0°	9.4130	9.4281	10.5719	9.9849	75.0°
15.1°	9.4158	9.4311	10.5689	9.9847	74.9°
15.2°	9.4186	9.4341	10.5659	9.9845	74.8°
15.3°	9.4214	9.4371	10.5629	9.9843	74.7°
15.4°	9.4242	9.4400	10.5600	9.9841	74.6°
15.5°	9.4269	9.4430	10.5570	9.9839	74.5°
15.6°	9.4296	9.4459	10.5541	9.9837	74.4°
15.7°	9.4323	9.4488	10.5512	9.9835	74.3°
15.8°	9.4350	9.4517	10.5483	9.9833	74.2°
15.9°	9.4377	9.4546	10.5454	9.9831	74.1°
16.0°	9.4403	9.4575	10.5425	9.9828	74.0°
16.1°	9.4430	9.4603	10.5397	9.9826	73.9°
16.2°	9.4456	9.4632	10.5368	9.9824	73.8°
16.3°	9.4482	9.4660	10.5340	9.9822	73.7°
16.4°	9.4508	9.4688	10.5312	9.9820	73.6°
16.5°	9.4533	9.4716	10.5284	9.9817	73.5°
16.6°	9.4559	9.4744	10.5256	9.9815	73.4°
16.7°	9.4584	9.4771	10.5229	9.9813	73.3°
16.8°	9.4609	9.4799	10.5201	9.9811	73.2°
16.9°	9.4634	9.4826	10.5174	9.9808	73.1°
17.0°	9.4659	9.4853	10.5147	9.9806	73.0°
17.1°	9.4684	9.4880	10.5120	9.9804	72.9°
17.2°	9.4709	9.4907	10.5093	9.9801	72.8°
17.3°	9.4733	9.4934	10.5066	9.9799	72.7°
17.4°	9.4757	9.4961	10.5039	9.9797	72.6°
17.5°	9.4781	9.4987	10.5013	9.9794	72.5°
17.6°	9.4805	9.5014	10.4986	9.9792	72.4°
17.7°	9.4829	9.5040	10.4960	9.9789	72.3°
17.8°	9.4853	9.5066	10.4934	9.9787	72.2°
17.9°	9.4876	9.5092	10.4908	9.9785	72.1°
18.0°	9.4900	9.5118	10.4882	9.9782	72.0°
18.1°	9.4923	9.5143	10.4857	9.9780	71.9°
18.2°	9.4946	9.5169	10.4831	9.9777	71.8°
18.3°	9.4969	9.5195	10.4805	9.9775	71.7°
18.4°	9.4992	9.5220	10.4780	9.9772	71.6°
18.5°	9.5015	9.5245	10.4755	9.9770	71.5°
18.6°	9.5037	9.5270	10.4730	9.9767	71.4°
18.7°	9.5060	9.5295	10.4705	9.9764	71.3°
18.8°	9.5082	9.5320	10.4680	9.9762	71.2°
18.9°	9.5104	9.5345	10.4655	9.9759	71.1°
19.0°	9.5126	9.5370	10.4630	9.9757	71.0°
19.1°	9.5148	9.5394	10.4606	9.9754	70.9°
19.2°	9.5170	9.5419	10.4581	9.9751	70.8°
19.3°	9.5192	9.5443	10.4557	9.9749	70.7°
19.4°	9.5213	9.5467	10.4533	9.9746	70.6°
19.5°	9.5235	9.5491	10.4509	9.9743	70.5°
19.6°	9.5256	9.5516	10.4484	9.9741	70.4°
19.7°	9.5278	9.5539	10.4461	9.9738	70.3°
19.8°	9.5299	9.5563	10.4437	9.9735	70.2°
19.9°	9.5320	9.5587	10.4413	9.9733	70.1°
20.0°	9.5341	9.5611	10.4389	9.9730	70.0°
	L cos	L cot	L tan	L sin	Degrees

TABLE 3 (Subtract 10 from Each Entry)

Degrees	L sin	L tan	L cot	L cos	
20.0°	9.5341	9.5611	10.4389	9.9730	70.0°
20.1°	9.5361	9.5634	10.4366	9.9727	69.9°
20.2°	9.5382	9.5658	10.4342	9.9724	69.8°
20.3°	9.5402	9.5681	10.4319	9.9722	69.7°
20.4°	9.5423	9.5704	10.4296	9.9719	69.6°
20.5°	9.5443	9.5727	10.4273	9.9716	69.5°
20.6°	9.5463	9.5750	10.4250	9.9713	69.4°
20.7°	9.5484	9.5773	10.4227	9.9710	69.3°
20.8°	9.5504	9.5796	10.4204	9.9707	69.2°
20.9°	9.5523	9.5819	10.4181	9.9704	69.1°
21.0°	9.5543	9.5842	10.4158	9.9702	69.0°
21.1°	9.5563	9.5864	10.4136	9.9699	68.9°
21.2°	9.5583	9.5887	10.4113	9.9696	68.8°
21.3°	9.5602	9.5909	10.4091	9.9693	68.7°
21.4°	9.5621	9.5932	10.4068	9.9690	68.6°
21.5°	9.5641	9.5954	10.4046	9.9687	68.5°
21.6°	9.5660	9.5976	10.4024	9.9684	68.4°
21.7°	9.5679	9.5998	10.4002	9.9681	68.3°
21.8°	9.5698	9.6020	10.3980	9.9678	68.2°
21.9°	9.5717	9.6042	10.3958	9.9675	68.1°
22.0°	9.5736	9.6064	10.3936	9.9672	68.0°
22.1°	9.5754	9.6086	10.3914	9.9669	67.9°
22.2°	9.5773	9.6108	10.3892	9.9666	67.8°
22.3°	9.5792	9.6129	10.3871	9.9662	67.7°
22.4°	9.5810	9.6151	10.3849	9.9659	67.6°
22.5°	9.5828	9.6172	10.3828	9.9656	67.5°
22.6°	9.5847	9.6194	10.3806	9.9653	67.4°
22.7°	9.5865	9.6215	10.3785	9.9650	67.3°
22.8°	9.5883	9.6236	10.3764	9.9647	67.2°
22.9°	9.5901	9.6257	10.3743	9.9643	67.1°
23.0°	9.5919	9.6279	10.3721	9.9640	67.0°
23.1°	9.5937	9.6300	10.3700	9.9637	66.9°
23.2°	9.5954	9.6321	10.3679	9.9634	66.8°
23.3°	9.5972	9.6341	10.3659	9.9631	66.7°
23.4°	9.5990	9.6362	10.3638	9.9627	66.6°
23.5°	9.6007	9.6383	10.3617	9.9624	66.5°
23.6°	9.6024	9.6404	10.3596	9.9621	66.4°
23.7°	9.6042	9.6424	10.3576	9.9617	66.3°
23.8°	9.6059	9.6445	10.3555	9.9614	66.2°
23.9°	9.6076	9.6465	10.3535	9.9611	66.1°
24.0°	9.6093	9.6486	10.3514	9.9607	66.0°
24.1°	9.6110	9.6506	10.3494	9.9604	65.9°
24.2°	9.6127	9.6527	10.3473	9.9601	65.8°
24.3°	9.6144	9.6547	10.3453	9.9597	65.7°
24.4°	9.6161	9.6567	10.3433	9.9594	65.6°
24.5°	9.6177	9.6587	10.3413	9.9590	65.5°
24.6°	9.6194	9.6607	10.3393	9.9587	65.4°
24.7°	9.6210	9.6627	10.3373	9.9583	65.3°
24.8°	9.6227	9.6647	10.3353	9.9580	65.2°
24.9°	9.6243	9.6667	10.3333	9.9576	65.1°
25.0°	9.6259	9.6687	10.3313	9.9573	65.0°
	L cos	L cot	L tan	L sin	Degrees

TABLE 3 (Subtract 10 from Each Entry)

Degrees	L sin	L tan	L cot	L cos	
25.0°	9.6259	9.6687	10.3313	9.9573	65.0°
25.1°	9.6276	9.6706	10.3294	9.9569	64.9°
25.2°	9.6292	9.6726	10.3274	9.9566	64.8°
25.3°	9.6308	9.6746	10.3254	9.9562	64.7°
25.4°	9.6324	9.6765	10.3235	9.9558	64.6°
25.5°	9.6340	9.6785	10.3215	9.9555	64.5°
25.6°	9.6356	9.6804	10.3196	9.9551	64.4°
25.7°	9.6371	9.6824	10.3176	9.9548	64.3°
25.8°	9.6387	9.6843	10.3157	9.9544	64.2°
25.9°	9.6403	9.6863	10.3137	9.9540	64.1°
26.0°	9.6418	9.6882	10.3118	9.9537	64.0°
26.1°	9.6434	9.6901	10.3099	9.9533	63.9°
26.2°	9.6449	9.6920	10.3080	9.9529	63.8°
26.3°	9.6465	9.6939	10.3061	9.9525	63.7°
26.4°	9.6480	9.6958	10.3042	9.9522	63.6°
26.5°	9.6495	9.6977	10.3023	9.9518	63.5°
26.6°	9.6510	9.6996	10.3004	9.9514	63.4°
26.7°	9.6526	9.7015	10.2985	9.9510	63.3°
26.8°	9.6541	9.7034	10.2966	9.9506	63.2°
26.9°	9.6556	9.7053	10.2947	9.9503	63.1°
27.0°	9.6570	9.7072	10.2928	9.9499	63.0°
27.1°	9.6585	9.7090	10.2910	9.9495	62.9°
27.2°	9.6600	9.7109	10.2891	9.9491	62.8°
27.3°	9.6615	9.7128	10.2872	9.9487	62.7°
27.4°	9.6629	9.7146	10.2854	9.9483	62.6°
27.5°	9.6644	9.7165	10.2835	9.9479	62.5°
27.6°	9.6659	9.7183	10.2817	9.9475	62.4°
27.7°	9.6673	9.7202	10.2798	9.9471	62.3°
27.8°	9.6687	9.7220	10.2780	9.9467	62.2°
27.9°	9.6702	9.7238	10.2762	9.9463	62.1°
28.0°	9.6716	9.7257	10.2743	9.9459	62.0°
28.1°	9.6730	9.7275	10.2725	9.9455	61.9°
28.2°	9.6744	9.7293	10.2707	9.9451	61.8°
28.3°	9.6759	9.7311	10.2689	9.9447	61.7°
28.4°	9.6773	9.7330	10.2670	9.9443	61.6°
28.5°	9.6787	9.7348	10.2652	9.9439	61.5°
28.6°	9.6801	9.7366	10.2634	9.9435	61.4°
28.7°	9.6814	9.7384	10.2616	9.9431	61.3°
28.8°	9.6828	9.7402	10.2598	9.9427	61.2°
28.9°	9.6842	9.7420	10.2580	9.9422	61.1°
29.0°	9.6856	9.7438	10.2562	9.9418	61.0°
29.1°	9.6869	9.7455	10.2545	9.9414	60.9°
29.2°	9.6883	9.7473	10.2527	9.9410	60.8°
29.3°	9.6896	9.7491	10.2509	9.9406	60.7°
29.4°	9.6910	9.7509	10.2491	9.9401	60.6°
29.5°	9.6923	9.7526	10.2474	9.9397	60.5°
29.6°	9.6937	9.7544	10.2456	9.9393	60.4°
29.7°	9.6950	9.7562	10.2438	9.9388	60.3°
29.8°	9.6963	9.7579	10.2421	9.9384	60.2°
29.9°	9.6977	9.7597	10.2403	9.9380	60.1°
30.0°	9.6990	9.7614	10.2386	9.9375	60.0°
	L cos	L cot	L tan	L sin	Degrees

TABLE 3 (Subtract 10 from Each Entry)

Degrees	L sin	L tan	L cot	L cos	
30.0°	9.6990	9.7614	10.2386	9.9375	60.0°
30.1°	9.7003	9.7632	10.2368	9.9371	59.9°
30.2°	9.7016	9.7649	10.2351	9.9367	59.8°
30.3°	9.7029	9.7667	10.2333	9.9362	59.7°
30.4°	9.7042	9.7684	10.2316	9.9358	59.6°
30.5°	9.7055	9.7701	10.2299	9.9353	59.5°
30.6°	9.7068	9.7719	10.2281	9.9349	59.4°
30.7°	9.7080	9.7736	10.2264	9.9344	59.3°
30.8°	9.7093	9.7753	10.2247	9.9340	59.2°
30.9°	9.7106	9.7771	10.2229	9.9335	59.1°
31.0°	9.7118	9.7788	10.2212	9.9331	59.0°
31.1°	9.7131	9.7805	10.2195	9.9326	58.9°
31.2°	9.7144	9.7822	10.2178	9.9322	58.8°
31.3°	9.7156	9.7839	10.2161	9.9317	58.7°
31.4°	9.7168	9.7856	10.2144	9.9312	58.6°
31.5°	9.7181	9.7873	10.2127	9.9308	58.5°
31.6°	9.7193	9.7890	10.2110	9.9303	58.4°
31.7°	9.7205	9.7907	10.2093	9.9298	58.3°
31.8°	9.7218	9.7924	10.2076	9.9294	58.2°
31.9°	9.7230	9.7941	10.2059	9.9289	58.1°
32.0°	9.7242	9.7958	10.2042	9.9284	58.0°
32.1°	9.7254	9.7975	10.2025	9.9279	57.9°
32.2°	9.7266	9.7992	10.2008	9.9275	57.8°
32.3°	9.7278	9.8008	10.1992	9.9270	57.7°
32.4°	9.7290	9.8025	10.1975	9.9265	57.6°
32.5°	9.7302	9.8042	10.1958	9.9260	57.5°
32.6°	9.7314	9.8059	10.1941	9.9255	57.4°
32.7°	9.7326	9.8075	10.1925	9.9251	57.3°
32.8°	9.7338	9.8092	10.1908	9.9246	57.2°
32.9°	9.7349	9.8109	10.1891	9.9241	57.1°
33.0°	9.7361	9.8125	10.1875	9.9236	57.0°
33.1°	9.7373	9.8142	10.1858	9.9231	56.9°
33.2°	9.7384	9.8158	10.1842	9.9226	56.8°
33.3°	9.7396	9.8175	10.1825	9.9221	56.7°
33.4°	9.7407	9.8191	10.1809	9.9216	56.6°
33.5°	9.7419	9.8208	10.1792	9.9211	56.5°
33.6°	9.7430	9.8224	10.1776	9.9206	56.4°
33.7°	9.7442	9.8241	10.1759	9.9201	56.3°
33.8°	9.7453	9.8257	10.1743	9.9196	56.2°
33.9°	9.7464	9.8274	10.1726	9.9191	56.1°
34.0°	9.7476	9.8290	10.1710	9.9186	56.0°
34.1°	9.7487	9.8306	10.1694	9.9181	55.9°
34.2°	9.7498	9.8323	10.1677	9.9175	55.8°
34.3°	9.7509	9.8339	10.1661	9.9170	55.7°
34.4°	9.7520	9.8355	10.1645	9.9165	55.6°
34.5°	9.7531	9.8371	10.1629	9.9160	55.5°
34.6°	9.7542	9.8388	10.1612	9.9155	55.4°
34.7°	9.7553	9.8404	10.1596	9.9149	55.3°
34.8°	9.7564	9.8420	10.1580	9.9144	55.2°
34.9°	9.7575	9.8436	10.1564	9.9139	55.1°
35.0°	9.7586	9.8452	10.1548	9.9134	55.0°
	L cos	L cot	L tan	L sin	Degrees

TABLE 3 (Subtract 10 from Each Entry)

Degrees	L sin	L tan	L cot	L cos	
35.0°	9.7586	9.8452	10.1548	9.9134	55.0°
35.1°	9.7597	9.8468	10.1532	9.9128	54.9°
35.2°	9.7607	9.8484	10.1516	9.9123	54.8°
35.3°	9.7618	9.8501	10.1499	9.9118	54.7°
35.4°	9.7629	9.8517	10.1483	9.9112	54.6°
35.5°	9.7640	9.8533	10.1467	9.9107	54.5°
35.6°	9.7650	9.8549	10.1451	9.9101	54.4°
35.7°	9.7661	9.8565	10.1435	9.9096	54.3°
35.8°	9.7671	9.8581	10.1419	9.9091	54.2°
35.9°	9.7682	9.8597	10.1403	9.9085	54.1°
36.0°	9.7692	9.8613	10.1387	9.9080	54.0°
36.1°	9.7703	9.8629	10.1371	9.9074	53.9°
36.2°	9.7713	9.8644	10.1356	9.9069	53.8°
36.3°	9.7723	9.8660	10.1340	9.9063	53.7°
36.4°	9.7734	9.8676	10.1324	9.9057	53.6°
36.5°	9.7744	9.8692	10.1308	9.9052	53.5°
36.6°	9.7754	9.8708	10.1292	9.9046	53.4°
36.7°	9.7764	9.8724	10.1276	9.9041	53.3°
36.8°	9.7774	9.8740	10.1260	9.9035	53.2°
36.9°	9.7785	9.8755	10.1245	9.9029	53.1°
37.0°	9.7795	9.8771	10.1229	9.9023	53.0°
37.1°	9.7805	9.8787	10.1213	9.9018	52.9°
37.2°	9.7815	9.8803	10.1197	9.9012	52.8°
37.3°	9.7825	9.8818	10.1182	9.9006	52.7°
37.4°	9.7835	9.8834	10.1166	9.9000	52.6°
37.5°	9.7844	9.8850	10.1150	9.8995	52.5°
37.6°	9.7854	9.8865	10.1135	9.8989	52.4°
37.7°	9.7864	9.8881	10.1119	9.8983	52.3°
37.8°	9.7874	9.8897	10.1103	9.8977	52.2°
37.9°	9.7884	9.8912	10.1088	9.8971	52.1°
38.0°	9.7893	9.8928	10.1072	9.8965	52.0°
38.1°	9.7903	9.8944	10.1056	9.8959	51.9°
38.2°	9.7913	9.8959	10.1041	9.8953	51.8°
38.3°	9.7922	9.8975	10.1025	9.8947	51.7°
38.4°	9.7932	9.8990	10.1010	9.8941	51.6°
38.5°	9.7941	9.9006	10.0994	9.8935	51.5°
38.6°	9.7951	9.9022	10.0978	9.8929	51.4°
38.7°	9.7960	9.9037	10.0963	9.8923	51.3°
38.8°	9.7970	9.9053	10.0947	9.8917	51.2°
38.9°	9.7979	9.9068	10.0932	9.8911	51.1°
39.0°	9.7989	9.9084	10.0916	9.8905	51.0°
39.1°	9.7998	9.9099	10.0901	9.8899	50.9°
39.2°	9.8007	9.9115	10.0885	9.8893	50.8°
39.3°	9.8017	9.9130	10.0870	9.8887	50.7°
39.4°	9.8026	9.9146	10.0854	9.8880	50.6°
39.5°	9.8035	9.9161	10.0839	9.8874	50.5°
39.6°	9.8044	9.9176	10.0824	9.8868	50.4°
39.7°	9.8053	9.9192	10.0808	9.8862	50.3°
39.8°	9.8063	9.9207	10.0793	9.8855	50.2°
39.9°	9.8072	9.9223	10.0777	9.8849	50.1°
40.0°	9.8081	9.9238	10.0762	9.8843	50.0°
	L cos	L cot	L tan	L sin	Degrees

TABLE 3 (Subtract 10 from Each Entry)

Degrees	L sin	L tan	L cot	L cos	
40.0°	9.8081	9.9238	10.0762	9.8843	50.0°
40.1°	9.8090	9.9254	10.0746	9.8836	49.9°
40.2°	9.8099	9.9269	10.0731	9.8830	49.8°
40.3°	9.8108	9.9284	10.0716	9.8823	49.7°
40.4°	9.8117	9.9300	10.0700	9.8817	49.6°
40.5°	9.8125	9.9315	10.0685	9.8810	49.5°
40.6°	9.8134	9.9330	10.0670	9.8804	49.4°
40.7°	9.8143	9.9346	10.0654	9.8797	49.3°
40.8°	9.8152	9.9361	10.0639	9.8791	49.2°
40.9°	9.8161	9.9376	10.0624	9.8784	49.1°
41.0°	9.8169	9.9392	10.0608	9.8778	49.0°
41.1°	9.8178	9.9407	10.0593	9.8771	48.9°
41.2°	9.8187	9.9422	10.0578	9.8765	48.8°
41.3°	9.8195	9.9438	10.0562	9.8758	48.7°
41.4°	9.8204	9.9453	10.0547	9.8751	48.6°
41.5°	9.8213	9.9468	10.0532	9.8745	48.5°
41.6°	9.8221	9.9483	10.0517	9.8738	48.4°
41.7°	9.8230	9.9499	10.0501	9.8731	48.3°
41.8°	9.8238	9.9514	10.0486	9.8724	48.2°
41.9°	9.8247	9.9529	10.0471	9.8718	48.1°
42.0°	9.8255	9.9544	10.0456	9.8711	48.0°
42.1°	9.8264	9.9560	10.0440	9.8704	47.9°
42.2°	9.8272	9.9575	10.0425	9.8697	47.8°
42.3°	9.8280	9.9590	10.0410	9.8690	47.7°
42.4°	9.8289	9.9605	10.0395	9.8683	47.6°
42.5°	9.8297	9.9621	10.0379	9.8676	47.5°
42.6°	9.8305	9.9636	10.0364	9.8669	47.4°
42.7°	9.8313	9.9651	10.0349	9.8662	47.3°
42.8°	9.8322	9.9666	10.0334	9.8655	47.2°
42.9°	9.8330	9.9681	10.0319	9.8648	47.1°
43.0°	9.8338	9.9697	10.0303	9.8641	47.0°
43.1°	9.8346	9.9712	10.0288	9.8634	46.9°
43.2°	9.8354	9.9727	10.0273	9.8627	46.8°
43.3°	9.8362	9.9742	10.0258	9.8620	46.7°
43.4°	9.8370	9.9757	10.0243	9.8613	46.6°
43.5°	9.8378	9.9772	10.0228	9.8606	46.5°
43.6°	9.8386	9.9788	10.0212	9.8598	46.4°
43.7°	9.8394	9.9803	10.0197	9.8591	46.3°
43.8°	9.8402	9.9818	10.0182	9.8584	46.2°
43.9°	9.8410	9.9833	10.0167	9.8577	46.1°
44.0°	9.8418	9.9848	10.0152	9.8569	46.0°
44.1°	9.8426	9.9864	10.0136	9.8562	45.9°
44.2°	9.8433	9.9879	10.0121	9.8555	45.8°
44.3°	9.8441	9.9894	10.0106	9.8547	45.7°
44.4°	9.8449	9.9909	10.0091	9.8540	45.6°
44.5°	9.8457	9.9924	10.0076	9.8532	45.5°
44.6°	9.8464	9.9939	10.0061	9.8525	45.4°
44.7°	9.8472	9.9955	10.0045	9.8517	45.3°
44.8°	9.8480	9.9970	10.0030	9.8510	45.2°
44.9°	9.8487	9.9985	10.0015	9.8502	45.1°
45.0°	9.8495	10.0000	10.0000	9.8495	45.0°
	L cos	L cot	L tan	L sin	Degrees

TABLE 4

No.	Square	Cube	Square Root	Cube Root	Reciprocal
1	1	1	1.0000	1.0000	1.000000000
2	4	8	1.4142	1.2599	.500000000
3	9	27	1.7321	1.4422	.333333333
4	16	64	2.0000	1.5874	.250000000
5	25	125	2.2361	1.7100	.200000000
6	36	216	2.4495	1.8171	.166666667
7	49	343	2.6458	1.9129	.142857143
8	64	512	2.8284	2.0000	.125000000
9	81	729	3.0000	2.0801	.111111111
10	100	1,000	3.1623	2.1544	.100000000
11	121	1,331	3.3166	2.2240	.090909091
12	144	1,728	3.4641	2.2894	.083333333
13	169	2,197	3.6056	2.3513	.076923077
14	196	2,744	3.7417	2.4101	.071428571
15	225	3,375	3.8730	2.4662	.066666667
16	256	4,096	4.0000	2.5198	.062500000
17	289	4,913	4.1231	2.5713	.058823529
18	324	5,832	4.2426	2.6207	.055555556
19	361	6,859	4.3589	2.6684	.052631579
20	400	8,000	4.4721	2.7144	.050000000
21	441	9,261	4.5826	2.7589	.047619048
22	484	10,648	4.6904	2.8020	.045454545
23	529	12,167	4.7958	2.8439	.043478261
24	576	13,824	4.8990	2.8845	.041666667
25	625	15,625	5.0000	2.9240	.040000000
26	676	17,576	5.0990	2.9625	.038461538
27	729	19,683	5.1962	3.0000	.037037037
28	784	21,952	5.2915	3.0366	.035714286
29	841	24,389	5.3852	3.0723	.034482759
30	900	27,000	5.4772	3.1072	.033333333
31	961	29,791	5.5678	3.1414	.032258065
32	1,024	32,768	5.6569	3.1748	.031250000
33	1,089	35,937	5.7446	3.2075	.030303030
34	1,156	39,304	5.8310	3.2396	.029411765
35	1,225	42,875	5.9161	3.2711	.028571429
36	1,296	46,656	6.0000	3.3019	.027777778
37	1,369	50,653	6.0828	3.3322	.027027027
38	1,444	54,872	6.1644	3.3620	.026315789
39	1,521	59,319	6.2450	3.3912	.025641026
40	1,600	64,000	6.3246	3.4200	.025000000
41	1,681	68,921	6.4031	3.4482	.024390244
42	1,764	74,088	6.4807	3.4760	.023809524
43	1,849	79,507	6.5574	3.5034	.023255814
44	1,936	85,184	6.6332	3.5303	.022727273
45	2,025	91,125	6.7082	3.5569	.022222222
46	2,116	97,336	6.7823	3.5830	.021739130
47	2,209	103,823	6.8557	3.6088	.021276596
48	2,304	110,592	6.9282	3.6342	.020833333
49	2,401	117,649	7.0000	3.6593	.020408163
50	2,500	125,000	7.0711	3.6840	.020000000

TABLE 4

No.	Square	Cube	Square Root	Cube Root	Reciprocal
51	2,601	132,651	7.1414	3.7084	.019607843
52	2,704	140,608	7.2111	3.7325	.019230769
53	2,809	148,877	7.2801	3.7563	.018867925
54	2,916	157,464	7.3485	3.7798	.018518519
55	3,025	166,375	7.4162	3.8030	.018181818
56	3,136	175,616	7.4833	3.8259	.017857143
57	3,249	185,193	7.5498	3.8485	.017543860
58	3,364	195,112	7.6158	3.8709	.017241379
59	3,481	205,379	7.6811	3.8930	.016949153
60	3,600	216,000	7.7460	3.9149	.016666667
61	3,721	226,981	7.8102	3.9365	.016393443
62	3,844	238,328	7.8740	3.9579	.016129032
63	3,969	250,047	7.9373	3.9791	.015873016
64	4,096	262,144	8.0000	4.0000	.015625000
65	4,225	274,625	8.0623	4.0207	.015384615
66	4,356	287,496	8.1240	4.0412	.015151515
67	4,489	300,763	8.1854	4.0615	.014925373
68	4,624	314,432	8.2462	4.0817	.014705882
69	4,761	328,509	8.3066	4.1016	.014492754
70	4,900	343,000	8.3666	4.1213	.014285714
71	5,041	357,911	8.4261	4.1408	.014084507
72	5,184	373,248	8.4853	4.1602	.013888889
73	5,329	389,017	8.5440	4.1793	.013698630
74	5,476	405,224	8.6023	4.1983	.013513514
75	5,625	421,875	8.6603	4.2172	.013333333
76	5,776	438,976	8.7178	4.2358	.013157895
77	5,929	456,533	8.7750	4.2543	.012987013
78	6,084	474,552	8.8318	4.2727	.012820513
79	6,241	493,039	8.8882	4.2908	.012658228
80	6,400	512,000	8.9443	4.3089	.012500000
81	6,561	531,441	9.0000	4.3267	.012345679
82	6,724	551,368	9.0554	4.3445	.012195122
83	6,889	571,787	9.1104	4.3621	.012048193
84	7,056	592,704	9.1652	4.3795	.011904762
85	7,225	614,125	9.2195	4.3968	.011764706
86	7,396	636,056	9.2736	4.4140	.011627907
87	7,569	658,503	9.3274	4.4310	.011494253
88	7,744	681,472	9.3808	4.4480	.011363636
89	7,921	704,969	9.4340	4.4647	.011235955
90	8,100	729,000	9.4868	4.4814	.011111111
91	8,281	753,571	9.5394	4.4979	.010989011
92	8,464	778,688	9.5917	4.5144	.010869565
93	8,649	804,357	9.6437	4.5307	.010752688
94	8,836	830,584	9.6954	4.5468	.010638298
95	9,025	857,375	9.7468	4.5629	.010526316
96	9,216	884,736	9.7980	4.5789	.010416667
97	9,409	912,673	9.8489	4.5947	.010309278
98	9,604	941,192	9.8995	4.6104	.010204082
99	9,801	970,299	9.9499	4.6261	.010101010
100	10,000	1,000,000	10.0000	4.6416	.010000000

INDEX

INDEX